THE HEIRLOOM GARDENER

THE HEIRLOOM GARDENER

Traditional Plants & Skills for the Modern World

JOHN FORTI

TIMBER PRESS 🐸 PORTLAND, OREGON

Published in 2021 by Timber Press, Inc.
The Haseltine Building
133 S.W. Second Avenue, Suite 450
Portland, Oregon 97204-3527
timberpress.com

Printed in China
Text and cover design by Sarah Crumb

ISBN 978-1-60469-993-7
Catalog records for this book are available from
the Library of Congress and the British Library.

To my many mentors,
and those who have touched me with
soulful landscapes and the spirit of place.

Contents

Preface

I COULD NEVER HAVE IMAGINED my career path as I wandered the riverside landscapes of my youth. My feet were on it, but I was too distracted by treetops and moccasin flowers to see it. Life in a small town (a great place to grow up "if you're a tree," according to my friends) quickly came to feel too small. Work at a garden center in my teens further ignited my interest in horticulture; it also helped me save up enough money to travel to Japan as an exchange student, far from my river and deep pine woods. There I saw the Japanese veneration of the land made manifest in regional artisanal foods, historic preservation, and the Zen-like devotion to the craft of gardening, the art of placing a single stone in a garden wall or a budding branch in an ikebana arrangement. I witnessed firsthand how much we are all shaped by place.

When I returned, I explored garden history and ethnobotany with deep interest. After college, I began working as a horticulturist for Plimoth Plantation, a living history museum, which grounded me in the native plants of the Wampanoag and the heirloom plants cultivated by the Pilgrims, the first immigrants to the region. I studied the craft of horticulture and the ethnobotany of early New England. Why did baskets, pottery, housing, and fashion take the shape they did? How did seasonal fluctuations and coastal living influence settlements, agriculture, diet, and spirituality?

A decade later I was recruited to become the curator of historic landscapes at Strawbery Banke Museum in Portsmouth, New Hampshire. On that ten-acre site, I worked to bring four centuries of original backyard gardens back to life in a waterfront neighborhood once slated for demolition in the name of urban renewal. My historical focus expanded from the 17th century to the examination of change, over time, to the cultural and historic landscapes of New England.

While there I helped to found and lead a chapter of the international Slow Food movement. We used the museum gardens and emerging farms to forge connectivity between the modern fabric of our community and the plants and historic foodways of early New England. Together with families, farmers, chefs, and brewers, we built a vital and vibrant model for a local economy that celebrates regional biodiversity and the interrelationship between community, human health, and the health and well-being of our shared environment.

In 2014, I became director of horticulture and education for the Massachusetts Horticultural Society. Once there, I opened the society grounds as a public garden. I built a collaboration between garden clubs, master gardeners, and regional farm-to-school programs to rebuild the school garden initiatives that the society had helped introduce to America in the 19th century. Perhaps most fun of all was getting to rehabilitate and create programming for a children's garden on the site.

Now I find myself living on the Maine shore of the Piscataqua River and working as the executive director of Bedrock Gardens in Lee, New Hampshire, an ingenious and artful landscape crafted from an old farm. In a world hungry for authenticity, it is an oasis of horticultural inspiration for new generations of visitors seeking serenity, sustenance, and meaning as we return to the land—with a passion.

My generation of Slow Food growers, thinkers, and collaborators understood that we were helping to create more sustainable and flavorful models for local food systems; we never imagined that we could rebuild local agricultural markets this quickly, or that a counter-revolution would be this necessary. "You never change things by fighting the existing reality," Buckminster Fuller once said. "To change something, build a new model that makes the existing model obsolete." Big ag may not yet be obsolete, but across the nation in little more than a decade, we have built and preserved meaningful alternatives.

Now, the urgent need for a more sustainable future is motivating new generations of home gardeners to reexamine what it means to craft a life in balance with community and environment. We are a post-industrial generation remembering the taste of quality food in a frenetic and fast-food world. We are seeking artisanal goods in a throwaway world, healing for a scarred environment. Our stewardship of the land and protection of our shared environment are at a crossroads. Fortunately, the way is clear to anyone with open eyes: sustainable practices can meet financial bottom lines *and* foster an environment that allows us to eat, drink, and breathe.

At a time when it would be easy to write about all that is wrong with the world, I write this as a celebration of the renaissance that is taking place in fields, backyards, and local economies around the world. This book is about building upon a sense of place to promote health, happiness, and common ground, whether it be for your own backyard homestead, farmstead, or community. I offer it as a garden historian's pathway to remembering the joys and lessons that pre-industrial technologies and heirloom garden crafts can offer if we choose to adapt them to foster a more sustainable future.

Introduction

HEIRLOOMS HOLD THE KEYS to the past, though their patina may have changed over time. They provide cultural context, common roots, and shared stories, and these are times to cultivate common ground. Heirlooms come in many forms, all of which involve a legacy of skill, craft, and memory. A cross-stitch sampler, a yellow-edged photograph, a recipe box or sourdough starter, a seasonal holiday tradition observed across generations. Unlike common antiques, heirlooms are artifacts of our life experience that beckon us to preserve them. Most come with a toolbox of clues that require our participation—an old jar of bean seeds calling us to plant them, a much-loved quilt in need of mending. When our hands pick up the work, they retrace life skills that, with a stitch in time, can help these timeworn artifacts live on for another generation.

With each stitch, each sprout, each taste, we unearth more of the world view, flavor, and history that encouraged people to save and pass along these heirlooms for another generation. In the process of becoming seed stewards, we can also learn to create our own future heirlooms. And have no doubt: thousands of preservation efforts like yours and mine all around the world can have a far greater impact than a global seed vault.

We live in a culture that seems to value us only as consumers. As such we are constantly reminded to avoid hassle and frustration and let the producers do the work for us. But to everything there is a season—and it may just be late autumn for the consumer culture craze. Younger generations don't want to shop 'til they drop. We *can* believe it's not butter, and we would rather buy from the farmer we know and help our kids understand where their food comes from. We recognize the toll that consumerism has taken on our health and environment. We know that pollution, species extinction, and climate change are real, and that we need to combine the latest science with traditional knowledge to move us into a more sustainable future.

Living high on the hog required a resource drain from society and the earth. In a consumer culture that can seem hollow and gutted, it is the process and community connection that makes for real and engaging experiences we could never purchase. My generation wants experiences: meeting friends at a farm wedding, watching seeds sprout, picking plants for dinner with our kids, brewing kom-

bucha, or making a wreath at a public garden. We want to build community gardens on our rooftops with neighbors and recycle tree limbs into garden fences and arbors over beers with a friend.

In a world full of plastic-wrapped lunchables and drive-through dinners, homegrown and family-prepared meals are a delicious act of resistance. Cultivating the patch of earth beneath our feet or trying our hand at garden craft fosters the stewardship and preservation of storied seeds and sustainable land-based skillsets—connections that help us remember how to build a stone garden bed, pickle produce, blend herbal teas, make salsa, or enjoy nutritious meals made with the fruits of our own labors. Head and heart, hand to hand. Connected to our history and landscape, across time, one seed at a time.

There are now as many people gardening as there were during the Victory Garden era of World War II, but our consumer culture ways have caused us to lose the craft behind cultivating our own seeds, grafting our own fruit trees, and building our own soil. Instead we buy lettuce in six packs, bags of "dirt" and fertilizer, and insta-garden pots with cherry tomatoes already on the vine. It's little wonder—that's the ethos we've been handed. But we're at a turning point, and the reality is that it's not just cheaper and more rewarding to know the process from seed to table, it's better for the environment, too.

Things like an old rhubarb patch, the remnants of an orchard, or a lichen-covered stone wall are talismans that help us read the landscapes we inherited. Through them, we catch a glimpse of how someone applied craftsmanship and the environmental arts to live in accord with nature. As heirloom gardeners in our shared backyard, we remember the work our hands were born to do, intuitively, like a bird follows its migratory path or a newly hatched turtle scrambles to the sea.

My goal is not to live in the past but to be present and participate in the betterment of my corner of the world. Experience my yard, my neighborhood, my watershed. The very parcel of land we dwell upon is an artifact, a living collection that, like an heirloom seed, requires our engagement; it is a diverse habitat, housing the microbes, fungi, bugs, birds, rabbits, foxes, herbs, and oaks that have roots in our plot of soil.

I may be a romantic, but I do not romanticize the past. In my work as a garden historian and herbalist, I am not blind to the shortcomings, biases, and errors of earlier times, but I also see families connected to seeds and soil, people connected to place, and a deep value for living in concert with our environment. In an age where traces of plastic and glyphosates are found in most of our foods, and landfills are brimming with excess, we know that we have lessons to learn about sustainable living. These discussions are best guided by a new generation of scientific and technological thinking that can help us to reexamine how we use and restore our land and to revitalize the horticultural, agricultural, culinary, and medicinal arts that guide us.

This book is an alphabetical collection of brief essays and artisanal images, each a seed, a way in to a different element of an heirloom gardening lifestyle; I see each entry as a point of connectivity—hand to hand, ancestor to descendant, seed to table. It's a love poem to the earth; a map to the art of living intentionally as stakeholders in a brighter future; a guidepost for gardeners, postindustrial makers, and explorers who want to cultivate common ground and craft new possibilities from local landscapes. I hope that this book inspires you to remember the craftsmanship behind building an heirloom garden and a happy homestead—that its pages provide not just survival and coping skills learned from the past but inspirations that rebuild community, restore and sustain our environment, and renew our quality of life through changing times. My wish is that art and craft will guide your hands to turn cabbage into sauerkraut, herbs into balms, and seeds into possibilities.

ANGELICA

A MAJESTIC HERB is *Angelica archangelica,* cultivated through the ages for its flavor, fragrance, and stately beauty. Seeds for this bodacious biennial were first brought to America in the 17th century, but they had a long history of use in Asia and Europe for medicinal preparations, teas, candies, and, famously, cordials: the emerging shoots epitomize the color chartreuse—so much so that they are the principle ingredient in the herbal liqueur of the same name.

In the garden, the hollow and resinous stems of this regal herb, covered in broad leaves, can easily tower three to five feet, and the enormous flower umbels rise up to seven feet toward the heavens—perhaps one of the reasons that the plant was dedicated to the archangels in Medieval times. Early each spring in centuries past, Europeans and Colonial Americans would harvest the tender stalks and simmer them in a simple syrup; eventually the stalks would become the translucent light green of sea glass, and the syrup would take on the color and herbaceous balsam flavor so unique to angelica. As lovers of spring have done long since, I repeat the process and candy the stalks until they become tender; I then either slice the stems lengthwise, into short segments, or braid the long strands together before rolling them in finely ground sugar to keep them from sticking to each other. They are excellent served like membrillo or marmalade

with cheese and dessert platters, and I end up using the shorter candied strips like citron in baking. Like an herbal equivalent to candied ginger, candied angelica was often served as a digestive at the end of feasts.

Throughout the growing season, but especially in spring and summer, I enjoy serving gin and tonics and other cocktails with straws made from thinner angelica stems. I also save the syrup that results from the candying process; it's an amazing herbal elixir to add into cocktails or serve atop vanilla ice cream.

To use angelica as a culinary herb, mince the tender leaves and add them to salads, cold fruit soups, and tabbouleh. As a tea herb, use whichever part of the plant is thriving: emerging leaves and stems in spring, flowers or tender seeds in summer, roots in fall and winter. In traditional Chinese medicine (TCM), the root of *Angelica sinensis* (dong quai), which has the same properties early American colonists ascribed to its cousin, is most commonly used for asthma and pneumonia. Both *A. archangelica* and *A. sinensis* form a great chartreuse backdrop for a perennial border or herb garden. *Angelica atropurpurea* and *A. gigas* offer the same but with purple-tinged aerial parts and deep purple umbels that are continuously abuzz with bees, butterflies, and birds. Like many of our lost or disappearing foods, angelica reminds us that many of the best plants for home gardeners may never make it to market in fresh form, so "grow it to preserve it" and enjoy!

APPLE CART

THROUGH the centuries, orchards and apple carts were symbols of the early American landscape, our agricultural marketplace, and sweet prosperity. Apples arrived in North America in the 17th century, brought as slips and seeds by the first immigrants. The first American apple was the Roxbury Russet, named in honor of the Massachusetts town where it grew.

Essential wisdom accumulates in the community much as fertility builds in the soil.

—WENDELL BERRY

Like many caught up in the same wave of immigration, my grandparents' arrival in the early 20th century came at a time when new industries were consolidating resources that could undercut and drive out centuries of local production. But knowing the ups and downs of the markets (and having seen many an apple cart topple), my grandfather bought a humble triple-decker in Roxbury as soon as he could. Like so many before them, my grandparents knew that the surest way to have food that was lovingly raised and affordable was to grow your own, and their courtyard garden became an Eden in the middle of a decaying city; several of their fruit trees stood on the same ground where a 17th-century governor kept his orchard.

At the same time, in the early 1900s, the temperance movement harnessed a religious fervor for sobriety that ushered in Prohibition. Not so ironically, Prohibition strategically killed local brewing and distillation, and powerful robber barons (who were the only ones able to work around the laws) stepped in to fill the void. These wily investors had bootleggers smuggle in liquor for their own speakeasies and private parties—where they no doubt sat back and made plans for the consolidated breweries and distilleries that would take over nationally once Prohibition ended. Meanwhile (file under "collateral damage"), our hard cider-drinking nation cut down huge swaths of ancient orchards in the name of Prohibition (followed by housing developments with names like Orchard Circle and McIntosh Lane). All lessons that we can learn from—or be doomed to repeat.

Today I see the apple cart as emblematic of our new local market economy. The return of local breweries and distilleries, and the revitalization of a regional landscape, complete with orchards. Farmers markets brimming with heritage apples in wooden crates, hard cider, and cider doughnuts. A cart maneuvered by a farming family in order to create a seasonal income stream directly between grower and eater. An alternative to plastic crates filled with machine-harvested, unripe apples gassed to turn red when they arrive at the distribution warehouse thousands of miles away. An apple cart without a half-dozen middlemen and fifteen added chemicals. Food justice on wheels ... and a chance to know the farmer and the stories of the produce behind the farm stand.

I have had the good fortune to partner with preservationists and foundations across the nation, working to save forgotten fruits (including regional apple varieties) that were disappearing at an alarming rate in a world of Red Delicious apples meant to endure shipping, look perfect, and remain shelf stable—at the cost of flavor and

regional diversity. Historically every state or county had a favorite apple adapted to the conditions and needs of the place: summer apples, storage apples, pie apples, sauce apples, and cider apples for every home and orchard. In my region, it was easier to grow apples than barley and more cost-effective to drink hard cider than beer; so now, with a return to local hard cider production, the apple cart even has a place out in front of our liquor stores.

I am grateful to see a younger generation tipping the apple cart and creating alternative systems. Red Delicious—thunk. High-spray orchards—dropped. Cargo ships laden with Chinese apples—sunk. Industrial juice—drained. In their place, pick-your-own heritage apples and pears, hard cider, and open spaces. Ka-ching: rebuilt local agricultural economies far more joyful and meaningful than data entry can collect. Systems which respect our shared cultural inheritance and invest in human industry.

The idiom to "upset the apple cart" has been with us, in some form, at least since Jeremy Belknap's *The History of New-Hampshire* (1784), which noted that John Adams nearly "overset the apple-cart" by slipping in his own amendment on the morning of the day the Constitution was to be ratified. Adams was a farmer-turned-politician and ultimately president at a pivotal time in our nation's history. Today, we are at another pivotal time, a time for the tenacity that gardeners know so well. It is clear that many of our systems are broken, but like other farmer-statesmen and -women before us, we are agitating, planting seeds, and rebuilding systems.

For those of us who love the land, there is a sense of urgency. It's time for us to get our hands back in the soil. Time for us to overturn politics as usual. We seem to be at a tipping point. Boom or bust. The bigger the mainstream gets, the more room there is for an undercurrent, and this generation of farmer-activists knows that some apple carts were meant to be overturned and filled anew—this time with local, organic, sustainably and justly raised heritage harvests, rich with flavor, history, and diversity.

ARTISANAL SKILLS

A NEW LAND-BASED ARTS and crafts movement is emerging in America. For more than a decade, local farmers and gig workers have been quietly reshaping the face of industry, and farmers markets and Etsy sites have grown up around them to create new retail outlets. In turn, these direct-to-consumer sales are reinvigorating cottage industries that produce small-batch quality goods for the sustainably minded: handcrafted wooden bowls, garden chutney, artisanal cheese, forged cutlery, and Windsor chairs. Cottage industries that lead to the preservation of a local mill, farmland, and heirloom seeds.

Much of the world seems to be coming around to the value of a new preservation movement, a responsive experience economy with both long-lasting and ephemeral goods that stretch old market perspectives on consumer culture. At this critical juncture, skill-based economies of scale remind us of things we should never have forgotten, basic skills that help artisans rebuild fractured economies and the environment, even as they regain a sense of pride in their work.

Historically, the barons of industry centralize power and make the cheapest product with the greatest profit margin, regardless of environmental or human impacts. But even as industry threatens to drown out our democracy, these new small-scale industries are proving to be a mighty undercurrent, offering consumers local artisanal

alternatives to unsustainable products sourced overseas to meet corporate bottom lines.

I never wanted to be defined as a consumer. I am a gardener and an artisanal producer, and I take care with the funds I expend. I want my home and garden to reflect William Morris's ethos: "Have nothing in your house that you do not know to be useful, or believe to be beautiful." Purchasing well-crafted household goods is part of a strategy to support alternatives to throwaway industrial manufactured goods. Instead of impusively buying into cheap particleboard that will end up in landfill, I wait until I can afford a locally crafted wooden table; meanwhile, I make more meals at home, supplemented with produce from my own garden and perhaps some fresh eggs from a friend. Being a producer in a gig household economy might mean managing a landscape to yield coppice for fencing, kindling, and charcoal, or making scratch bread or kimchi for home or for sale. Artisanal work in my community runs the gamut from compost contractor to local greengrocer; it's the local distiller who turned around production to make hand sanitizer during the pandemic, and the farmer in the adjacent field who grows the grapes and grains that supplied him.

Fortunately, many former boomer-consumers and hippies, who saw an earlier back-to-the-land movement, are now in a position to support sustainable systems. In my region, that has meant community-minded entrepreneurs investing in outdoor infrastructure for year-round open markets, participating in the revitalization of local agriculture, and organizing regional distribution hubs that make it more lucrative for those willing to wholesale directly to alternative food chains. Alternative is frequently lambasted as elitist by the consumer culture that helped get us into this mess, but to me it means my middle-class grandmother's thrift foods, made with whole

A man who works with his hands is
a laborer. A man who works with his
hands and his head is a craftsman;
but a man who works with his hands
and his head and his heart is an artist.

—LOUIS NIZER

ingredients from her own city backyard or sources she otherwise knew. It means that eggs with deep yellow yolks, crisp cucumbers, and fresh sausage fill our corner stores—local alternatives to soda and Ring Dings.

Technology. The word is derived from the Greek, *techne* ("skill," "craft") and *logos* ("word," "thought")—in essence, a personal expression of skill made manifest. I like to think that sustainable technologies are once again united with the crafts that gave birth to local industries. An alternative to life as a disconnected cog in a manufacturing wheel, or a market demographic attached to a blue screen. For many of us, our garden is both a personal artistic expression and an applied craft that keeps us grounded in meaningful work.

It's been suggested that the difference between art and craft was that craft was women's work or manual labor; or worse, that both art and craft are matters of novelty—this from the generation that reduced them to painted polyester aprons and crocheted Kleenex-box covers. But look back further, to trade unions or guilds: artisanal work was highly valued, as it came with an assurance of quality. Fortunately, many of the new generation of young farmers and artisans are women and men who blend the integrity of guilds with flavorful and useful products that inspire fair recompense.

Craft industries are akin to public utilities, which, by their nature, should do the important work of preserving our shared natural resources. Artisanal skills can help us rebuild our common wealth by producing sustainable, place-based ingredients, nontoxic waste, and regenerative agriculture. We can create compost instead of landfills, farm food over factory food. Fewer plastics, more parsnips.

"If the robot makes it, let the robot buy it" is a sentiment posted by the register at many of the finer establishments in my town. Our community works to buy products made by the people and industries in our shared landscape, and the quality of life here is better for it. In an age of robotics and artificial intelligence, artisanal skillsets provide meaningful alternatives to soulless mass production. Machines don't know what to do with a ripe heirloom tomato or an oddly shaped potato. Machines don't understand human dignity, pride, or individuality. Machines require standardization, which has cost the world precious variables, including biodiversity, sustainability, and humanity. You can't stamp out craftsmanship on a production line. It's one of the many reasons why so many of us bypass the robot: it doesn't fit our values stream, and it seldom serves those living downstream.

Modern industry often disrespects human dignity and the planet; and without the union wages, insurance benefits, and pensions that built the Greatest Generation, new generations are opting out. They are adapting pre-industrial technologies and agricultural skills to foster new economies where we can buy from each other. Beer and butter, farm shares or a timber-frame shed. Veg for vegans. Maple syrup for tree work. Cottage industries bring us goats for chèvre, meat, and poison ivy control. They bring honey for mead and art for art's sake. Apprenticeships and jobs linked to the land bring honor and meaning back to manual labor in our gig economy, and cash on the barrel head. No gym membership necessary.

When we produce, sell, and buy in our own communities, seventy percent of money spent remains in our local economy; the new baker in town is commissioning farm and garden production of grains, seeds, nuts, fruit, and botanicals, a graphic designer, a media sales rep, an electrician, a carpenter, a local mill, a community kitchen space, and perhaps even apprentices who help to bring goods to market or open up a bakery storefront café. We become conscious of our kith and kin—our friends, acquaintances, and relations,

those we are glad to support and those we may not be familiar with yet. The average child in America knows fewer than ten backyard plants and animals; perhaps we need to reintroduce them to their kinfolk: the toad and salamander, pipsissewa and plantain. Kin that sting, kin that eat tomato hornworms, kin that pollinate the food we eat. Artisanal work can help rebuild these communities of kith and kin, which can occur any-where—an agricultural landscape, sea-side communities, urban and suburban
yards. When we participate, we reclaim an alternative to the impersonal, the robot-made. In the process, we can craft local alternatives to preserve the cultural inheritance of our kith and kin in a garden right outside our own door.

I have greater faith in the future when I see local schools placing orders with nearby farmers, and see the kids touring those same farms, where they learn firsthand about an aquifer, a turbine, and life in a colony of bees. I take hope from a chef friend who bought up a local farmer's entire hail-damaged harvest to make frozen soups for the school system instead of watching a neighbor suffer the loss of an annual crop. I saw my community buoyed as a local farm distribution hub adapted services to deliver vital spring produce to a population in pandemic isolation. Local industries keep us resilient when we watch systems "too big to fail" fail.

In many ways, we are rebuilding from the ground up. The richer we make the soil, the more meaningful the harvest. Gardeners live artful lives. Each year that we cultivate, the earth teaches us more about the skills to craft a life and livelihood from seeds and stems. We learn of the harmony that comes from turning barren earth into a colorful and fruitful artist's palette. Each season, our roots and desires to live sustainably inform design inspirations framed by the natural world around us—from luscious spring green through harshest winter gray. Apple cider to zinnias.

Garden historian Mac Griswold had it right: "Gardens are the slowest of the performing arts." As performing artists, every pot of herbal tea we make is artisanal, a blend of art, craft, and science. Each garden bed, the artistic vision of our inner painter, chef, and perfumer. Each compost pile, the artist's medium, the activist's plan of action, and a sandbox for kid engagement—and rogue pumpkins. Gardens are bringing back time for reflection and connection. They can offer a snail-paced still life that helps us to savor the fruits and flowers of our accomplishments and remember the gardeners who taught us, even if only for a few sacred moments in a busy week.

Artisanal skills can provide a good meal or a livelihood. They offer a form of play and innovation that keeps us entertained and intellectually challenged throughout the seasons of our life. Artisanal living gives us an opportunity to vote with our forks by supporting the farms, gardens, and agritourism sites that crop up in its wake, not the strip malls, processing plants, and food deserts that sadly remain too prevalent in corporatized dystopian landscapes. Artisans skillfully dovetail hands and heart—and as all gardeners know, that can surely heal the earth, even if only one backyard at a time.

BEE BALM

BEE BALM IS A FRAGRANT native herb with bright tubular flowers full of citrusy floral nectar. It is a boon for bees and hummingbirds and a sweet, soothing balm for families with frayed nerves. Fortunately, it is also perfect for inclusion in herbal teas, salads, and cocktails.

The generic name of this tenacious sun-loving perennial honors Spanish physician and botanist Nicolás Monardes, whose *Joyfull Newes Out of the Newe Founde Worlde* (1577), which extolled the virtues of bee balm and other New World plants, encouraged settlement in the Americas. After nearly two hundred years of the same settlers overlooking indigenous plants and importing familiar plants from Europe, we had our first native plants movement, at the outset of the American Revolution. Colonists, intent on cutting ties with Great Britain, banned imported black teas and designated bee balm as one of the premier Liberty Teas (substitutes for black tea); it was a highly effective embargo that ultimately helped to win the war with balms instead of bombs. By the Victorian era, bee balm had become our homegrown alternative to the Mediterranean citrus bergamot, the chief flavor note in Earl Grey tea.

In the garden, bee balm is a perennial favorite and airy spark of color. As with so many plants, its common name suggests its use: today, we recognize bee balm as a key plant for reviving declining populations of bees and other pollinators. We also know this resinous

plant as a balm, a fragrant ointment used to heal or soothe the skin and beleaguered spirits. When I don't have access to the fresh plant, I sprinkle the essential oil around my house, or put a drop on my pulse points when I need a lift. All aerial parts of the plant can be used fresh. The tender leaves can be chopped finely and added to salads and summer vegetable dishes, or made into teas and balms. However, the flowers (particularly the red ones) are my favorite edible part of the plant; they are full of an Earl Grey–like floral nectar and offer a fun kid candy from the garden. They are beautiful frozen into ice cubes for summer drinks, or turned into a simple syrup for refreshing cocktails and confections. *Monarda didyma* is the red-flowering South American type documented by Monardes; its mauve-flowering cousin, *M. fistulosa*, is a North American native. Both species are quick to spread via underground runners, so I will often fill teapots with the thinnings I pull up from the ground to keep it contained.

Many patented native cultivars have been selected or developed to reduce height, primarily in order to fit on the cramped shipping carts used to supply box stores; others are bred to be more resistant to powdery mildew. However, this Yankee organic gardener and citizen scientist has long noted that bees have a strong tendency to ignore plants that have been sprayed with growth inhibitors. Furthermore, I have yet to see a plant (even those inclined toward powdery mildew) develop a chronic case of that fungal disease if top watering is avoided; it also helps to apply a generous two- to three-inch mulch of pine needles to the bed where they grow prior to the start of the season. Above all, I have seen monarda work as a magic balm when engaging children in gardens. There is great reward in teaching kids to sit still with a flower in hand, long enough to be visited by a hummingbird. There is an equal joy to make time to do the same for ourselves.

BIODIVERSITY

When we
try to pick out
anything by
itself, we find
it hitched to
everything else
in the Universe.

—JOHN MUIR

BIOLOGICAL diversity is the sum total of all living organisms that exist on our planet, the remarkable variability among organisms and ecosystems that comes down to us from hundreds of millions of years of evolutionary history. We can see it in humans, but many people have difficulty distinguishing variables in the earth beneath their feet, and the wall of green they see when they look at nature. Genetic diversity among plants is critical, for the same reasons that captive animal breeding programs need to introduce new mates from wild populations. Species with wide genetic diversity tend to produce a wider variety of offspring, some of them potentially the most fit variants. In contrast, a species with little or no genetic diversity will produce offspring that are genetically alike and therefore more susceptible to diseases.

Similar circumstances occur in the plant world. The potato famine that drove nearly two million people to flee Ireland, my paternal grandmother among them, may be traced to the lack of genetic diversity among potatoes. There was only one "lumper," and when it became blighted, all its genetic clones did, too. A lesson we would have done well to learn from. Yet instead, over the last hundred years, we have lost more than ninety percent of the genetic diversity among crops that fed the world. A recent United Nations biodiversity report confirmed that species loss is due to human activity, and scientists report that we are losing almost two hundred species per day. We are in the midst of a sixth mass extinction.

Biodiversity can seem daunting, beyond us as caring gardeners, parents, educators, and consumers to mend; yet collectively, we are equipped to be a successful worldwide frontline to rebuild genetic and ecosystem diversity. One yard at a time. One community at a time.

Many among us were raised with an inheritance of garden stories and recipes. The genesis of these stories is kept alive in heirloom seeds, which remind us all of who we are and where we come from. The skills that meant we could feed ourselves throughout the seasons are increasingly important if we want to ensure good nutrition and food security in this problematic world.

The U.S. Department of Agriculture considers fruit and vegetables specialty crops; corn, wheat, soy, and rice are considered commodity crops. American agribusiness values commodity crops more because they are publicly traded on a commodities market and most require processing—two things that serve corporations but not the health of the environment, our health, or biodiversity. Farmers that receive subsidies for growing commodity crops are even prevented by law from growing any fruit or vegetables—or diversity—in their landscape.

Historically, we knew over twenty thousand plants that could be eaten or used for textiles, cordage, or medicinal, cosmetic, and creative purposes. Every county had a favorite bean, early berry, or nutritious salad green that extended the growing season; but most of that was lost when we allowed chemical companies to hybridize and patent the seeds handed down to us as our nation began corporatizing agriculture. Today the world primarily relies on about two dozen commodity crops that appear, ad nauseam, on grocery market shelves.

We all know that specialty (fruit and veg) crops are healthier for us, yet commodity crops have literally been shoved down our throats. Sixty-eight percent of Americans understand that organic is healthier for them and the planet, but less than one percent of farming in America is organic, and organic products represent only four percent of goods sold. It's why our gardens and family farms are more important than ever, and it's why organic is the largest-growing market share. Unfortunately, we cannot realistically shift conventional agriculture without a major overhaul to agribusiness, or the reintroduction of a nationwide network of biodiverse organic farms. And the latter is where I am seeing the greatest (agri)cultural shift.

Gardeners and farmers know that when our plantings are diverse and the soil is rich, the worm is fat—and the early bird gets the worm. The flock eats the grubs, borers, and caterpillars, and the corn grows more disease resistant, yielding a healthier crop. A healthy crop benefits all our beneficial co-conspirators, from the microbes in healthy soils to birds, bees, and small family farmers.

Gardeners also know that the food on our plate can become one of the key ways by which we support biodiversity. Gardening inspires many of us to cook instead of purchasing processed foods, which most often rely on the lowest common denominator of ingredients. Produce like the russet potato, 08B2084 sweet corn, and Cavendish banana are not so much bred for consumption as genetically engineered to be uniformly shaped, mechanically harvested, and capable of enduring several thousand shipping miles before arriving

at market. Cultivating heirlooms and supporting small, diverse agricultural enterprises (instead of vast monocultural conglomerates) are two ways we can help to revitalize ancient grains, digestibility, flavor, crops that self-fertilize, and plants adaptable to a changing world.

Fortunately, seeds remember better than we do. Seed saving is an ancient garden craft that engages each of us as backyard preservationists, working to improve our lot with each passing season and every generation. We have long understood that if we saved the seed of our best produce, it would broaden biodiversity and continue to increase yield, disease resistance, and vigor.

Conventional crops call upon farmers to use chemical fertilizers and pesticides to contend with the difficulties that arise when you grow contiguous miles of the same agricultural crop in the same place, year after year. Remedies that are generally worse than the cure. Treatments that kill every living organism in the field—good, bad, or indifferent.

Every year, gardeners have a fresh chance to become backyard heroes and heroines working to preserve the regional success stories embedded in the plants our ancestors used to feed themselves and the world. In the process of caring, we become global participants in a movement to restore some of the diversity lost over the last century of big ag. As we establish community and school gardens, we can link up new fertile green corridors that help to restore genetic and ecosystem diversity, much as Victory Gardens during World War II helped feed our nation and improve homefront morale. Fortunately, we are once again seeing successful regional models crop up again from coast to coast. Many remind us that planting flowers is essential for the mutual success of our pollinators. Many rekindle local agriculture and foodways by embracing the Slow Food movement, the basis for building a delicious revolution—one by which we relearn how to garden, farm, forage, cook, and preserve from a place of diversity. A new revolution that comes from a joyful place of discovery and reconnection to community and landscape.

We are not obliged to live with systems that harm biodiversity. The overwhelming majority of the population does not want GMOs. It's time to turn the compost pile. If we are unable to access the organic food we want, we must become part of the undercurrent. Use the language of history, indigenous knowledge, and the wisdom of our kith and kin from the natural world to say *enough*. We know a better way. We are defending the natural world that supports all life.

Biodiversity speaks in many languages. Bumblebee and Bangladeshi bitter melon. Zebra and zucchini. Languages that speak of earth as our mother, oak trees as brothers, and rivers as veins that course with life. Kin that require respect for all they do to enrich our habitat, nourish our bodies, and replenish our spirits. I am happier to anthropomorphize and elevate my fellow citizens of the earth—whether green, scaled, feathered, or furry—and the web we share than to grant a corporation personhood. I am more comfortable trusting my gardener's instincts for living in a reciprocal relationship with the land than I am trusting a corporation to "do the right thing."

In our own lifetimes, environmental degradation and a loss of genetic diversity has meant the disappearance of bugs on our windshields, whip-poor-wills on our windowsills, and unmown fields and roadsides teeming with native grasses, milkweed, autumn asters—and life. It has narrowed the selection of foods on our plate; caused food allergies, diabetes, and bee colony collapse; and devastated habitat for monarch butterflies and other bellwether species worldwide. Every species lost is another broken strand in the web that supports us.

The voices of diversity all need to be heard in solidarity. Voices of women, children, indigenous, elderly, minorities, and the impoverished. Biodiversity brings breadth and depth to our conversation. It helps to foster interdisciplinary explorations that can open up new pathways and creative solutions. Creativity, inspired by evolution and adaptation, holds the keys. Even the best bioengineers can only manipulate genetic codes. Nature created them, and biodiversity programmed the key code.

Gardeners know that the green path is medicine for our mind, body, and spirit. We can't forget the skills, seeds, and kinship to be learned from our ragged history—the good, the bad, and the ugly. All can guide us forward as we work to regenerate landscapes that support critical biodiversity.

Too often these days, building a house means overbuilding a lot. We strip every tree from the landscape. We scrape away the native soil and every sign of life. Every microbe. Mycelia and morels. Flickers and partridgeberries. We erase all evidence of nature simply because we had the tools to dominate it. Simply because we enable the contractors, insurance agents, and land speculators to dictate how it's done, instead of making it a crime against nature. So many of our technologies can be tied back to weapons of war, and now we turn them against the earth. Fortunately, biodiversity offers a counterbalance to the monocultural landscapes of recent years. A refreshing movement is afoot to renounce the ten billion blades of identical grass that make up each of the estimated forty to fifty million acres of lawn in America. A shift that is taking root, one tiny patch of clover, daisy, and meadow at a time.

A fresh look at biodiversity is helping motivate us to resurrect local farms that cultivate a diversity of crops. These local farms are helping to lead the way back to operating without the chemical fertilizers and pesticides that are killing off pollinators and destroying habitat.

If society declares that corporations should be afforded the rights of personhood, we should do the same for our environment and all the diversity contained within it. Clearly corporations can and do take care of themselves, but it is equally clear that fields, fisheries, and rainforests also need investment. Fortunately, the easiest place to start is in our own backyard compost pile.

BOTANIZING &
HERBARIA

DID YOU HAVE A TEACHER, relative, or friend who taught you to love nature? Perhaps someone who taught you the languages of your environment, like plant names, bird calls, animal tracks, celestial bodies, and cloud formations?

Cloud means something different today, and internet screens are crafted to reel us in. The recognition of nature-deficit disorder, introduced by Richard Louv in 2005, turned into a call to action to help balance out our children's time. A call to get kids back outside and create meaningful legislation to provide equal access to our shared resources. But we are not the first generation called to reengage kids in the outdoors. As we shifted from an agrarian society to an industrial nation, Americans saw waves of children working in mills and factories; in response, many Victorians took to botanizing—the study of plants in their natural habitat—in order to deepen connections to the natural world. With a 19th-century passion for fresh air, crafts, and science, parents and children headed to the woods and meadows with sketch pads, books of identification, magnifying glasses, and a flask of switchel lemonade.

Hand in hand they documented the flora where they lived. They pressed botanical specimens in herbaria, wrote all the names and identifiers they knew, and documented cultural conditions to blend art and science in the field and on the page. They romped through woodlands to identify lady's slipper orchids and pipsissewa. On their way home they picked from the seasonal bounty, foraging for fiddleheads, raspberries, and hickory nuts; once there, they sipped wintergreen tea and made terraria, berry bowls, and herbal crafts with what they had collected. Expedition field days were in fact fun science lessons from nature's classroom which helped families learn and remember household botany together.

To further educate children about local plants, kids were often taught how to craft their own herbarium. Herbaria are collections of pressed and dried botanicals made into reference books. They included details like cultural notes, common and scientific names, culinary and medicinal attributes, and the date and location where the specimen was gathered. They were used as personal plant identification manuals, derived from the student's life experience and environment; and they taught science and stewardship to any child with a naturally inquisitive mind. They also became an artful component of botanizing that paired creativity and lifelong learning in a personalized document of place and time.

Herbaria served as scientific worksheets for young household botanists. Kids adapted the skills learned to dry and press flowers for decorative arts and crafts, including bookmarks, tea trays and tables with specimens under glass, and framed floral art to hang in their homes. Young women would keep herbaria as part of an ongoing life lesson, similar to the way in which they would stitch a sampler as an expression of the accumulated skills they had learned (alphabet, numbers, sewing, embroidery). Herbaria were also kept by doctors, pharmacists, and explorers dedicated to scientific analysis of medicine and the natural world. The ones kept by Emily Dickinson,

Henry David Thoreau, and others help us to understand environ--
mental stewardship more deeply today; by analyzing bloom times,
regional plant communities, plant dispersal, and other valuable
long-term data, we can better document the impacts of plants and
climate change.

Today, herbaria continue to offer great interdisciplinary science
lessons from the schoolyard or home garden. And making an herbarium remains one of my favorite crafts for engagement in the natural
world; it's a fun do-it-yourself way for you and your family to share
time and explore nature together. At any time in life, if you want to
make your own herbarium, just follow this simple guide:

1. Find a botanical specimen.
2. Cut or pull your specimen from a place where you know it is
   abundant.
3. Try to collect as much of the plant as possible (roots, shoots,
   leaf, stem, flower, seed).
4. Identify your plant (use a field guide with keys, send a photo to
   an expert, or look online).
5. Press and dry your specimen.
6. Attach it to an herbarium sheet with paste, ribbon, or narrow
   strips of tape.
7. Write the common and scientific names of the specimen along
   with the date/location where it was found, and any other details
   that will help you remember the plant and the excursion years
   from now.

If a child is to keep his inborn sense of wonder, he needs the companionship of at least one adult who can share it, rediscovering with him the joy, excitement, and mystery of the world we live in.

—RACHEL CARSON

Maybe someday your herbarium will become a family heirloom, or an ongoing history of place in a university or museum!

In addition to spending time in nature, botanizing reinforced the skills of household botany and the domestic arts, which modern culture has abandoned in the age of consumerism. It helped people create place-based food, beverage, and medicine along with ephemera like May baskets and holiday wreaths. Inadvertently, botanizing also taught us to create tangible heirloom keepsakes, a process that enabled each maker to elevate craft to a personal expression of art, whether as a sketchbook, an embroidered flower, a collection of recipes, or an herbarium. For those that endured the ages, they linger like a flavor, forever associated with the time and place of the maker.

At the historical height of botanizing, many with the means traveled great distances, not only across North America but to Europe and Asia, but most people engaged just to notice and celebrate the first fragrant mayflower or the autumnal combination of aster and goldenrod as they walked home from school or factory work. Regardless, nature does far more than simply provide us with food, shelter, and data. It civilizes us, too, inspiring people to create a national park system that preserved places like the Grand Canyon from greed and exploitation. These Victorian botanizers also raised a generation wise enough to tend Victory Gardens and put by the harvest before those skills were lost.

When we pause to reconnect with the cycles of nature—maybe even long enough that we forget to recharge the laptop—we remember how important it is to disconnect. I don't worry too much about the next generation. I watched too much TV as a kid (and I don't even own one as an adult), but along the way, anyone can be taught to love and respect nature.

My first nature teacher was my aunt Claire, a child of the 1920s with her head in the '60s. She emulated Saint Francis, and her love of nature was infectious. She taught me to walk silently, "like an Indian," and to hold still long enough for a chickadee to land on me. Together we would lie flat on pine needles, surrounded by ostrich fern, looking up to see the shapes of clouds; we examined skunk cabbage melting snow as it rose through the ground in the dell where springs flowed to the river. We identified catkins and bird calls and raised countless generations of orphaned squirrels, chipmunks, and fledglings until they were old enough to release back into the wild. I still think of my aunt Claire and smile, knowing that I owe her a debt of gratitude for helping to raise me in habitat instead of fear.

I was also drawn to nature by the magic revealed daily in the schoolyard by our science teacher, Miss Parsons, who encouraged us to explore a fascinating world I hadn't yet realized was my own. I learned that respect and curiosity were the keys to wandering in that realm, and once that door opened, I came to know the feeling of cool moss underfoot, the snap of wild asparagus, and the tang of sheep sorrel in my mouth. I learned the secret ways of snails, the coded language of wrens, and—with a little help from my flora and fauna friends—the art of coexistence.

Maybe your naturalist elders are still living, but Aunt Claire and Miss Parsons are gone, and we are the rising generation's mentors now. We need to share our love of the environment with our kids, grandkids, neighbors, and communities. We need to hit the reset button and take a walk where we live. Who are you mentoring? Think back to your favorite experiences in nature, or the foraging walk you took with your mentor, and head outside to botanize with a young one. You will remember enough. The rest will follow.

CHESTNUT

I'VE CARRIED A CHESTNUT in my pocket since I was in my early twenties. It was already worn to a polish when it was given to me by an old man who had saved the glossy brown nut from the last American chestnut tree he had known. It sparked my interest and offered smooth, round hope—of good luck for a young man and an old tree.

In living memory, chestnuts made up twenty to seventy percent of the forests in parts of North America. Our native American chestnut (*Castanea dentata*) grew to be an enormous and stately tree in the forest canopy. It provided an exceptionally hard wood for barns, log cabins, house sills, fences, furniture, railroad ties, and telephone poles. The nuts were a significant part of the diet for turkey, deer, and other woodland wildlife. Unlike acorns, which require special preparation to make them edible for humans, the American chestnut was naturally sweet and the only chestnut in the world that could be eaten without being roasted first. For this reason, they figured large in the diets of Native Americans and generations of immigrants alike, as free foraged protein. For centuries, whole towns would go out "chestnutting" for harvest festivals and holiday celebrations. Like seasonally foraged Maine blueberries, Cape Cod cranberries, or mushrooms, the nuts contributed significantly to rural and household economies; they were so abundant in autumn that farmers would herd hogs and cattle to chestnut-dominated forests

to sweeten and fatten them before butchering them for home use or for market. Chestnut ripening coincided with the holiday season, from Thanksgiving to New Year. Hot roasted chestnuts were a favorite winter market offering; they could be bought from street vendors and enjoyed as handwarmers and snacks for caroling and hearthside enjoyment. Turn-of-the-century newspaper articles often showed train cars overflowing with chestnuts rolling into major cities to be sold fresh or roasted.

All this changed with *Cryphonectria parasitica*, a pathogenic fungus thought to have been accidentally introduced from imported nursery stock in 1904. What came to be known as chestnut blight was first found in chestnut trees on the grounds of the Bronx Zoo. By 1940, most mature American chestnut trees had been wiped out by the disease. It is estimated that at that time there were four billion American chestnut trees, yielding an annual crop of twenty million pounds in the eastern United States. But no more. When it can be found, the hardwood is so enduring that skilled artisans across the nation still recycle old fence rails, barnboards, and timbers to make fine furnishings, tools, and household items.

At Plimoth Plantation, my first horticultural stint after college, I curated exhibits, living history interpretations, and craft and living collections for both the Wampanoag and Pilgrim sites. In the process of recreating the gardens and landscapes of the 17th-century inhabitants, I would research the native and heirloom plants known to be grown and used during that era. When primary sources guided me to the American chestnut, I planted six disease-resistant trees from the American Chestnut Foundation throughout the grounds; these trees are hybrids, a cross between trees that survived the chestnut blight and just enough Chinese chestnut (*Castanea mollissima*) to make them disease resistant.

Every child should have mud pies, grasshoppers, water bugs, tadpoles, frogs, mud turtles, elderberries, wild strawberries, acorns, chestnuts, trees to climb. Brooks to wade, water lilies, woodchucks, bats, bees, butterflies, various animals to pet, hayfields, pine-cones, rocks to roll, sand, snakes, huckleberries and hornets; and any child who has been deprived of these has been deprived of the best part of education.

—LUTHER BURBANK

A few years ago, I returned to the museum as an invited speaker. After my talk, I walked down to the Wampanoag homesite. The warm welcome there overwhelmed me—until I was distracted by the sight of an unfamiliar tree canopy on the horizon. It was one of the disease-resistant chestnuts I had planted a decade earlier—and it was full of kids! My first thought was to holler, like my grandfather might have, "You, kids—get down out of that tree!" But as I walked closer, I choked back my words and silently wiped tears from my eyes instead. Half a dozen Wampanoag kids were playing in the tree and collecting nuts, likely the first Native American children to harvest chestnuts in that region in nearly a century. Another chestnut, one I'd planted at the visitor center (to honor the 1987 visit of Prince Akihito and Princess Michiko of Japan), now stood taller than the roof of the building—a testament to how quickly trees grow, the good work of the American Chestnut Foundation, and how swiftly the years pass.

Later, as curator of historic landscapes at Strawbery Banke Museum, I reintroduced a pair of American chestnut seedlings that came from a rare surviving tree across the river in Maine. Each tree planting represents a scientific experiment, an act of preservation, a shady canopy, and a culinary delight. In Portsmouth, I am still Mr. Castagno (Italian for "chestnut") to an old man who knew me for roasting chestnuts at the museum's annual candlelight stroll. There I would work with volunteers to cut an X (I can hear my Italian grandfather's voice saying "a cross") in each chestnut, so they wouldn't explode when we began to roast them. Throughout the night, visitors and carolers had edible handwarmers in their pockets and a song on their lips: *Chestnuts roasting on an open fire.* All year long, discarded shells cast onto the curbs and sidewalks of the neighborhood served as reminders that chestnuts have been a celebratory witness to history. And if we plant more trees, we sow a better future.

CORDIALS

OUR COUGH SYRUPS AND EXPECTORANTS are mostly the post-Prohibition stepchildren of their botanical antecedents. Cherry, ginger, and elderberry linger on in name, but once Prohibition divorced alcohol from medicine, chemicals and sweet syrups replicating efficacious herbs became carriers instead of the alcohol, which formerly preserved and enhanced the somnorific effect of healing herbs. In Europe the concept of food and drink as medicine better survived attempts at prohibition and the pressures of a pharmaceutical industry. Drinks served before a meal to stimulate appetite (apéritifs in France, aperitivos in Italy) are typically dry and lower in alcohol, though a cocktail can count as an apéritif, too.

A cordial (digestif in France, digestivo in Italy) is meant to stimulate or support digestion or overall good health. The term harkens back to the time of the early English immigrants, who referred to any sweetened syrup or medicine that used sugar, honey, maple syrup, or molasses for a preservative as a cordial. In the United States, cordial is used interchangeably with liqueur, although it is most often associated with a sweet, sipping drink to be enjoyed with cheese, fruit, nuts, or prepared desserts.

Where I live, cordials have a long history of preserving the flavors and medicinal attributes of botanicals gathered ripe from the sun. They are typically made by steeping or distilling digestive (bitter) herbs, or fruits and spices, with sugar and alcohol. In the middle of winter, I can taste the sun-saturated ripeness and healing benefits of these liqueurs, which are intended to heal everything from digestive disorders to coughs, colds, and the winter blues.

Like tinctures, cordials were a way of preserving medicinal herbs, fruits, roots, barks, flowers, and seeds in alcohol—but with enough sweetness to offer a sipping medicine that would warm mind, body, and spirit. Historically, they were prescribed for cough, colds, and digestion; and as we now know, many of the herbs and fruits used in cordials are indeed rich in antioxidants and vitamin C and are powerfully effective digestives, cough suppressants, and expectorants.

Some cordials are distilled, but in my home apothecary, I typically fill a glass jar or carafe two-thirds of the way with my favorite, very ripe botanicals and cover them completely with whatever base alcohol seems to pair best—most often that's high-proof, unflavored vodka, brandy, gin, or rum. After two to three days in the sun, set the concoction aside in a dark, room-temperature space for one to three months—how long will depend on how quickly the botanicals infuse flavor. Once they have released enough flavor to tantalize your taste buds, strain out the botanicals and sweeten the resulting liqueur to taste.

Having lived through many a bitterly cold New England winter, I can lend firsthand testimony and praise for a cordial's warming effect. Yankees may not be famous for their warmth, but they knew enough to self-prescribe a dose from an elegant decanter to warm the cockles of the heart (as my Irish grandmother would say) and to help make the cold winter, or a group gathered for the holidays, more cordial.

DANDELION

> A weed is a plant that has mastered every survival skill except for learning how to grow in rows.
>
> —DOUG LARSON

LINNAEUS originally (1753) placed dandelion in the genus *Leontodon* (from the Greek, "lion's tooth"), a reference to the plant's deeply serrated leaves. Lion's teeth is still a common name for the dandelion in many languages, including French, where the *dent de lion* (corrupted to our "dandy lion") is also *pissenlit* (the common colonial-era translation of which was "piss-a-bed"), at once recognizing the plant's diuretic properties and forewarning parents not to serve it to young ones in suppertime salads.

Dandelions arrived in North America in the 17th century. According to most herbals of the time (and many present-day researchers), they were introduced by colonial gardeners with intention, not as weed seed: dandelions were frequently added to early spring salads (though not at bedtime) and used to stimulate digestion and support liver and kidney function.

In the Victorian age, sharp spiraling blades replaced the brushes in carpet sweepers, and the lawn mower was created. Just as quickly, the idealized monocultural American lawn was born—a sign of the rising leisure class and the framework for a new suburban landscape.

Garden books, ladies' magazines, and seed catalogs of the time offered idyllic images of lawns with children rolling hoops, playing croquet, and botanizing along the margins. Post–Civil War marketing for the fashionable suburban landscapes shifted from work yards and kitchen gardens to more "civilized" grounds that were punctuated by vast carpet beds of exotic annuals that a new greenhouse trade made possible. The foundries that had made armaments during the war years turned around production to make the first generation of garden furniture, urns, pools, and gazebos, which were coveted by the newly minted middle classes. A lawn symbolized rest and recreation for those who aspired to join the ranks, and there was little room for the once-beloved humble dandelion in this new green framework.

Not everyone was swayed by the allure of these new monocultures. The journals of Sarah Goodwin, the wife of New Hampshire's Civil War governor, were the basis for the gardens we recreated at Strawbery Banke. They provided numerous garden and landscape insights, but they also spoke beautifully to the shift towards uniformity she was witnessing and critiquing when she declared, "I do not allow either buttercups or dandelions to be dug up." They lived on in her lawn because she still saw beauty and purpose in them.

Despite such lamentations, dandelions have been targeted by the marketing arm of our lawn "care" industry for over one hundred years. Instead of food, medicine, and cheery yellow flowers for pollinators, dandelions were quickly consigned to the category of deeply hated spectacle and much-maligned lawn weed. They must be controlled (if we have any hope of being good neighbors), and so we buy into a costly lawn care regime.

When I established a Victorian children's garden at Strawbery Banke, in a site adjacent to Sarah Goodwin's garden, one of the spaces I created was a zoo border, complete with 19th-century images of animals the plants were named for. The dandelion patch was labeled with an illustration of a roaring circus lion, and the words "Dandelion—from the French Dent de Lion—or Teeth of the Lion." Yet volunteers and visitors couldn't resist "weeding" the patch of dandelions every time we replaced them. Ironically, tenacious weeds don't conform well, and our transplants were frequently doomed to failure, while just beyond the fence, their wild cousins were growing through sidewalk cracks.

A bright young intern who helped install the children's garden taught me to make fritters from dandelion flowers; each spring, I am reminded of her when I make them. Growing up, we didn't have many dandelions flowers to make fritters from our lawn. Not because of weed killers—unless my Italian grandmother's passion for the bitter herbs of early spring made her a weed killer. Each year, before the yellow flowers rose up, my grandmother would get out to harvest dandelion greens from the lawn with her penknife. She would sauté them with a little olive oil, minced garlic or onion, and a squeeze of lemon, dash of vinegar, or fistful of diced tomato, and we would all enjoy a side dish of spring together. She would also dice up and dry the root for liver and kidney teas and tonics to be used later in the season. I keep those traditions alive each spring, but now I also enjoy eating the unopened buds, or sautéing and watching them open in the eggs I scramble for a little springtime alchemy and a silly dose of garden magic.

Today we recognize dandelions as part of the symphony of life in our landscapes. We understand that, as an important food source for early pollinators, they should be left in place when we work to create pollinator-friendly habitat areas, including golf courses, campuses, civic buildings, and backyards, all working in concert to shift the paradigm for ornamental landscaping in America.

DISTILLATION

DISTILLATION GETS DOWN to the root of the matter, reducing abundant garden botanicals to precious essential waters and oils. It is the essence of a botanical rendered into marvelous things: booze, perfume, medicine—and so much flavor. Mention distillation, and for many people the first image that comes to mind is someone making moonshine in a still, deep in hill country. But for an heirloom gardener, a still can open up a panoply of distilled delights, made from the natural world.

Stills first entered my gardening life at Plimoth Plantation, when I found "helm" listed in the 17th-century inventory of Captain Myles Standish. I was clueless as to what a helm might be—and thus began my descent down a rabbit hole that has turned into a twenty-plus-year passion for distillation. Evidently, this early style of still was called a helm because it resembled a military helmet of the day—basically, an upside-down pot that you might have seen the cast of Monty Python wearing in very silly skits. It's a simple design, but one that captures the essence of distillation.

In its most rudimentary form, distillation is like boiling water for broccoli, but instead of the condensation running back down the lid into the pot, it is carried off via a condenser coil to render ... well, essence of broccoli. Fortunately, early apothecaries and perfumers

had higher aspirations. Most distillation in Medieval Europe took place in monasteries, where the monks made medicines and (medicinal, ahem) liqueurs for themselves and the wider community. When the English church broke from Rome and monasteries were handed over to sympathetic nobles, many gentlewomen took to stillhouses like she-sheds—no doubt a welcome retreat for those who were still considered chattel, but also a creative outlet for the gardener who furnished her outbuilding not only with the tools of her craft but bundles of fragrant herbs and a still. Quite often, noblewomen who took over these monastic gardens continued making and distributing medicines. Anyone who has ever distilled knows that it takes a huge volume of botanicals to make an essential oil or booze. But high-style gardens of the day were large and often included knot gardens with intertwining herbal hedges of lavender, rosemary, santolina, hyssop, germander, southernwood, savory, and sage; and groundcovers of fragrant dianthus, marjoram, chamomile, violets, heartsease, sweet woodruff, strawberry, and thyme—all herbs that needed to be trimmed regularly to conform to hedging, or to fill the still. So, many of these landed ladies took to processing each season's harvest into medicines, perfumes, and liqueurs.

In recent decades, aromatherapy finally provided a name for the long-held understanding that fragrances can shift our physical and emotional state of well-being. Early herbalists, who knew that not all medicines needed to be ingested to be alterative, made notes like this: "To smell a rose brings a smile to the face, thus the virtue of a rose is to gladden the heart." To gardeners, who understood subtlety and fragrance, this rang true as they extracted medicinal virtues and scents from plants.

Hydrosol is the aromatic water that remains after steam-distilling botanical materials. When a rose is distilled, a great deal of hydrosol or rosewater carries away the dilute essence of the flower, but since oil and water don't mix, rose (or any other botanical) oil floats to the surface. These essential oils are so called because they were said to represent the very essence of botanical fragrance and flavor.

When plants like rose, lavender, or orange blossom are distilled they are often referred to as floral or toilet waters (from the French *toilette*—and the 18th-century English word for the room where the grooming, washing, and perfuming occurred). Aromatic waters have properties similar to essential oils but are much less concentrated. They usually have a scent similar to their essential oil, perhaps with a greener note, and a shorter shelf life (the water-soluble constituents in plant material are not present in the essential oil). Old distillation bottles often had a pinch in the neck that would allow the essential oil to form an airlock to preserve the more volatile waters below.

Hydrosols were typically used or sold as affordable byproducts of distillation: lightly fragrant perfumes, flavorings, cleansing agents, astringents, wash waters, and insect repellents. Witchhazel and orris root water are still distilled for use as astringents. Lavender water (beneficial in repelling moths and bedbugs) carries a long history of use as a laundry water for washing linens; and its antiseptic floral scent proved useful as a toilet water well appreciated in an age without much access to hot water or propensity for bathing.

Rosewater may be the most beloved essence used throughout the centuries and across cultures. Both roses and distillation arrived in English gardens from the Middle and Far East, and with stills came a passion for rosewater. It was used in cosmetics and as an astringent facial skin toner, wash water, and all-around courtship refresher; but once wed and bed, everyone loved rosewater best in the kitchen. Much as it is still in the Middle East, rosewater was used in baked goods, fruit pies, candies, confections, and in nearly every way we think to use vanilla today. In fact, it persisted in Old and New World cuisine until the 19th century, when vanilla extract (made by tincturing vanilla orchid seedpods) became all the rage.

The essential oils and liqueurs extracted in the process of distillation carry with them the very same chemical components that helped the living botanical attract pollinators or protect itself from predators. The same phytochemicals that made plants antimicrobial, working to the benefit of the plant, could be distilled down into potent medicines, beneficial cosmetics, fragrant perfumes, essential flavorings… and yes, a tasty digestif, aka booze! Liqueurs were sipped (held under and swirled around the tongue to hasten absorption into the bloodstream) in order to aid digestion, calm nerves, and induce sleep. Essential oils would be used on pulse points (neck, behind ears, wrists), their scent activated with each beat of the heart and its warm pulse. They were so strong, one little dab and your treatment was complete—or your lure was set.

My own distillation journey began when two friends who were buyers for the museum shop where I worked haggled on my behalf for a cool old copper still and the classic pinched-neck bottles that came with it. The antique dealer said it had belonged to a Cape Cod doctor; it looked to date from the late 18th or early 19th century, as did the glass bottles. Some antiques look fragile and call us to preserve them on a shelf, but these were as sturdy as an old ax ready to fell another tree. My first experiment, during rose season, yielded fine rosewater and a few precious drops of essential oil. From there it was on to lavender, mint, and any other botanical I could harvest in sufficient quantity.

When I had progressed as much as I could alone, I invited distillation maven Jeanne Rose to Plimoth Plantation as a speaker. She had been a leader in the American herbalism revival that began in the 1960s, but on this visit, she brought her in-depth knowledge of perfume: high and low notes, floral, earthy, woody, herbaceous—our heads were spinning. When she returned the next summer to conduct an intensive program on distillation and perfumery, it felt like I'd gone from cranking a Model T to high-speed rail. As a young apprentice, I was grateful to find a wise mentor who had such deep experience with an ancient garden craft.

By the time I turned forty, I was in pursuit of another still—this time in Turkey, one of the historic hotbeds for distillation. I traveled far and wide, trying to explain what I was looking for. I knew the old name for a still was alembic (or limbeck), but what I didn't know until my last week there was that if you stressed the last syllable, more like the Old French *alambic* (ahl-em-beek), or better yet, referred to the product of the still, the Arabic *alkuhul* (al-co-hol!), a far better result ensued. I was immediately hastened across town to the apothecary district, where to my surprise, my only options were finely tempered scientific glass alembics. Not what I expected to leave the ancient city with, but a fine floral still nonetheless. I flew home with the still in a box (labeled "alembic") on my lap. No questions asked.

A few years later—at a Slow Food summit in Italy, in an enormous hall of vendors selling rare and endangered foods and agricultural products—I found my workhorse: a brand-new copper still made by a family that had been keeping the tradition alive for seven generations, My first chance to use the new still was at Strawbery Banke's annual candlelight stroll. Ordinarily, I would teach wreath making and roast chestnuts, but that year I obtained a large harvest of wintergreen from a farmer friend. Soon the minty scent of wintergreen, perfect for the holidays, filled the entire historic house, and for the next few nights, the distillate was used to make old-style wintergreen candies and candy canes.

These days, my still of choice is larger and of a classic gourd-like design—though it does gall me somewhat that after going across the world, it was found on the internet. As a rule, I don't buy things online, but this ideal alembic was the exception that proves the rule. Every time I use it, I am reminded that we are all outgrowths of terroir and distillates of our ancestors and experiences, the essence of history, place—and a little bit of alchemy. And sometimes all it takes is a breath of sweet perfume, and I am carried off on distant journeys once more.

DOORYARD GARDENS

AS EARLY AS THE 18TH CENTURY, we found middle ground between the utilitarian kitchen gardens behind our houses and the showier spaces shared with passersby. The dooryard garden was a streetside opportunity to put one's best foot forward, with the most elegant plants, pathways, and high-style fences that could be afforded. Unless the house was too close to the road, the dooryard offered enough space to supply the household with bouquets of fragrant flowers or dainty dishes of ripe berries that would never make it to hot summer markets. Like boundary markers for the property, shade trees (most commonly native sugar maples, later elms in urban areas) lined roads, providing fall color and sweet maple syrup. Perennials, collected from neighboring woodlands, friends, and gardens around the world, adorned the edges of pathways and occasionally accented foundations. Flowers like daylily, peony, and iris were often shared from grandmothers' gardens to those of new brides. The corners of the house were most frequently planted with lilac, as a harbinger of spring and standard bearer for fragrance; elder, for its medicinal properties and beauty; quince, for its elegant flowers and fragrant fruits; or American holly, for year-round interest.

In sunny entryway landscapes, thyme covered the pathways, and borders were lined with fragrant herbs (dianthus, marjoram, lavender, southernwood) or perennials. If the dooryard garden was shady, Solomon's seal, sweet cicely, sweet woodruff, foxglove, ostrich fern, mayapple, bleeding heart, violets, and lily of the valley grew as companions along the brick, cobble, or gravel paths. Where there was space for an arbor, climbing roses, grapes, morning glory, or other fruiting or flowering vines would provide additional shade, fragrance, and beauty. Adjacent fruiting beds often contained edible currants and gooseberries, rhubarb, and strawberries; secondary paths or flower borders, yucca, poppies, hollyhock, comfrey, gentian, and mallows; and high walls or fences might even have espaliered fruits trained upon them.

What's old is new again. I still like my backyard privacy, but dooryard gardens put an outward face on the finer gardens we keep. They provide a point of contact with the community—something sorely needed in these days of neighbors pulling up driveways and into the garage, never to be seen again. As we seek to minimize lawn and find sunny spaces to grow meaningful plants, as we come together to find common ground and a new face for American landscapes of the 21st century, it may just be time to convert some of that lawn and your overgrown hedge into a new version of the dooryard garden. Besides curb appeal, it can offer a fruitful intersection from which we can once again get to know our neighbors—sharing plants, skills, beauty, and conversation.

EDIBLE FLOWERS

IF YOU ARE WHAT YOU EAT, why not eat beautiful things? Old herbals suggested that flowers held the sweetest essence of the plant, and they recommended eating them just as often as they noted the benefits of leaving them for pollinators and seed saving. The romantic in me is pleased to take inspiration from these early gardeners, grateful to know that we could use flowers for more than perfumery and bouquets. Bees, butterflies, and hummingbirds know this too; the tubular flowers of bee balm and pineapple sage are full of nectar waiting to be sipped as an aid to pollination. Yet they are just as sweet for us to eat, sprinkled on a salad, or to drink in a cup of tea.

Common herbs like mint, marjoram, chives, basil, cilantro (coriander), and thyme have edible flowers that mimic the herbal savor of the plant with a distinctive sweetness added by the nectar they contain. Try mint flowers instead of leaves in tabbouleh; sprinkle them on fruit salad, or fresh cut slices of melon. Fresh mint or rosemary flowers also yield a refined cup of tea; the difference is subtle but elegant.

We go to great lengths to dress up a salad of tasteless iceberg lettuce with tomatoes, regardless of season, when one made from the flowering tops of herbs and perennials can turn a salad into a celebration of wellness, beauty, and seasonality. In early spring—violets,

heartsease, pansy, sweet cicely, Solomon's seal, and chervil. In early summer—roses, lavender, dianthus, borage, bee balm, sage, and daylily. In high summer and autumn—nasturtium, calendula, fennel, dill, and pineapple sage. These, along with other tender leaves and shoots of the herbs and veggies we grow, can make any salad a homegrown work of art, even if you are just adding some backyard or windowbox diversity to market greens.

Some restaurants, trying to be trendy, buy and serve bland orchids, clumsy carnations, and dense daisies—lumpish flowers more akin to a tasteless winter strawberry with scarcely a blush of red. While they might look novel, these flowers offer little taste or flavor. True gardeners, however, have the distinct advantage of gathering in flowers sweet with nectar at the height of flavor.

When our gardens are at their peak, ripeness and simplicity can be all it takes to make a meal. May wine steeped with the flowers of sweet woodruff, angelica, Solomon's seal, and violets. Dandelion or elderflower fritters. Slices of vine-ripened tomato strewn with basil flowers. Cucumbers dusted with the pungent flowers of dill and a splash of chive blossom vinegar. Fresh fish fillets dusted with the flower and pollen of fennel. Juicy peaches halved and bedecked with bee balm. Herb butter blended with sage flowers or calendula and a hint of honey, and even the flowering tips of bay leaf and rosemary as a wintry warmer in a teapot.

Edible flowers don't have a long shelf life, so they don't generally find space on our supermarket shelves. As a result, most of us have forgotten flavors familiar to virtually every generation of gardeners before us. I see edible flowers not as a fancy, frilly addition to food but simply as a home gardener's opportunity to use every—and often the very best—part of a plant, and one that is typically is untapped. Most importantly, they provide a chance to make a beautiful and nutritious salad that is also the bee's knees.

ELDER·

MORE AND MORE, AS AMERICANS relearn the value of fresh, local foods, foraging is in the news. Often when we imagine people gathering wild plants, we think of hardy souls harvesting precious delicacies in distant woodlands. Yet many of my favorite native fruits and gourmet greens are toughing it out, right in our own urban land-scapes. Elder is one such plant. Its fountain-like growth habit and conspicuous spring flowers help even the novice to recognize its presence in a surrounding landscape.

I first learned about elder from downriver neighbors, who stirred my childhood imagination with tales of Native Americans and 17th-century settlers using it for food and medicine. Many of the arrowheads we would find along the river originally had shafts made from the stems of elder, and early settlers did indeed apply their knowledge of the European *Sambucus nigra* to the use of our native *S. canadensis*. The resulting elderberry cordial would have had pride of place in a colonial medicine cupboard or in a glass decanter behind the bar of a tavern. Today, more prosaically, I most often see elder's large, linen-white umbels along roadsides and highways, wav-ing like flags in the breeze.

Over the years, I have distilled the flowers to make elderflower water and liqueur, fried elderflower fritters, put up elderberry wine, and turned the berries into vinegar, jam, and syrup. The syrup has become a favorite addition to seltzer water and cocktails, or simply drizzled atop ice cream. At the height of the season, I make several variations of rhubarb or mixed fruit and elderberry pie, and once I collaborated with a local brewery to create a fruit gruit, a Medieval-style ale with elderberries and herbs. Of all that I do with elder, I would have to say that infusing the berries in vodka with honey, sugar, or best of all, syrup of elderberry, makes my favorite "cordial medicine" or

elderberry liqueur. Its antioxidant, expectorant, and ascorbic properties may well help with respiratory ailments and strengthening the immune system in winter, but it also lives up to its earlier name as a medicine that will make us more cordial.

Elder has a long history of use on this continent, and in the homelands of many of the first immigrants to arrive on these shores. When we work with rare and underutilized native or heirloom produce like elder, we are renewing North American culinary traditions. We are also applying an ethnobotanical approach to the study of plants, which helps us to meld folkloric knowledge of health-giving benefits with 21st-century science. As in the past, we take elderberry as a flavorful and healthy expectorant medicine, but now we know that the fruit also provides ascorbic acid. Somehow, First Peoples and early settlers knew to ward off scurvy and winter colds with it, long before they were aware of vitamin C, or thought to import orange juice.

Many interesting cultivars of elder are available, but try to plant the straight native *Sambucus canadensis*, so that the birds and critters you share it with are able to spread nonsterile, indigenous seeds back into the environment. Elderberries are deciduous shrubs, fun and easy to grow, and they make a beautiful landscape plant, some five to twelve feet tall at maturity. Plants thrive in average to damp ground, in part to full sun. Showy fragrant flower umbels open in June and July, attracting bees and butterflies and, by midsummer, foraging birds, foodies, and "fathers who smell of elderberries."

Bringing history to life from our gardens not only keeps many artisanal traditions alive, but it helps to form better historical perspective and a delicious sense of place. We are not only connecting with the plants, flavors, and recipes of those who were here before us; we are rebuilding biodiversity and a cultural memory that teaches us how people managed to stay healthy and feed themselves affordably, year-round, in a local economy. If you want to preserve heirloom plants for the future, the best thing to do is to grow and eat them.

ELM

FROM EARLIEST TIMES, Native Americans traversed the inland waterways and coastlines of eastern North America in dugout canoes made from the water-resistant trunks of American elm (*Ulmus americana*). Early settlers steamed and bent the hard wood of this valuable lumber tree to make barrel and wheel hoops, boxes, rocking chairs, and baskets. Later it was used to make other furniture, veneer, and even hockey sticks (for those with pent-up energy in winter). One of the museums where I worked had several hollowed logs in its collection, fashioned from rot-resistant elms; they had conveyed water into homes throughout the town for hundreds of years.

Elms reached the height of their popularity in the 19th century, when unsustainable agricultural practices had degraded the landscape, leaving it nearly barren of trees. Mill culture, railroads, and spent soil induced a rural flight from beleaguered farms into cities and their factories. For many, it seemed easier to abandon agriculture and ship agricultural products back from the Midwest via railroad and depot stations. As we abandoned the wide-open, sun-baked East Coast farms for the fertile Midwest, American elms became the shade tree of choice and, within decades, the unparalleled street tree for newly minted suburban and urban America. We lined our avenues with them to the exclusion of almost every other tree.

The elm's stately vase-shaped canopy created tall verdant tunnels that sometimes continued for miles in an otherwise treeless landscape of former fields. They provided fast-growing shade for home and marketplace, magnificent cool corridors that drank up puddles and brought birds, soil, and life back into the landscape. Elm trees were the first blush of green in the new generation of parks and city streetscapes emerging nationwide.

Unfortunately, as history has shown us repeatedly, too much of a good thing comes at a cost. In the early 1930s, when Dutch elm disease crept in, it was able to wipe out our native elm because there was no diversity. The fungal disease was carried by elm bark beetles, and it spread like wildfire through the monocultural street-lining canopy. These trees proved as susceptible to an imported disease as Native Americans had been to the illnesses of earliest European colonists. They are a potent reminder of why we need to plant for genetic diversity. When there is diversity, there will always be resilient survivors.

Some say that tradition is just peer pressure from our ancestors. Elms reminds us that not every tradition is worth handing down. Many people still lament the disappearance of streets lined with American elm, but we have learned the hard way that having miles of the same "crop" comes at an environmental cost.

Today, many elm restoration projects around the country are working with diverse American elm tree strains that have high levels of tolerance to Dutch elm disease. The process of regeneration will allow the American elm to co-evolve with the Dutch elm fungal pathogen, ensuring that this valuable tree species will not be entirely lost from the American landscape. Fortunately, not even these organizations recommend recreating the monocultural street plantings of the past; rather, they recommend botanical moderation and regeneration.

Over the past twenty years, I have planted many of these disease-resistant elms. Some of them are already more than fifty feet tall, providing shade and shelter for all manner of life. I plant them like some might plant sugar maples or ginkgo. Some accent streetside corners of public buildings; others provide shade for historic homes and wigwams. At my own home, I have nurtured a native seedling that desperately wanted to grow out of the perennial border along the street. Now some forty feet tall, it casts summer shade for my house—and the long shadow of a noble history.

ETHNOBOTANY

DOES YOUR FAMILY HAVE plants with powerful associations? An herbal tea you used for medicine? A vegetable prepared or preserved with traditional methods, or a plant used ceremonially? An ethnobotanist would be keen to hear all about it. Ethnobotany is the study of a region's plants and their practical uses through the traditional knowledge of a local culture. Most often these days we employ ethnobotany to look abroad for plant-based medicines that our pharmaceutical companies replicate and patent chemically. We would learn a great deal more by applying the principles of ethnobotany to history, using each season's harvest to restore historical horticultural practices, to better understand how earlier residents of a given place used plants for textiles, construction, food, medicine, and spirituality.

Whether you are a scientist working for a pharmaceutical company, a DIY crafter exploring how to make twine from milkweed on YouTube, or a mixologist recreating an old recipe for haymaker's punch, ethnobotany offers a deep dive into our connections with the natural world. Similarly, the soil we work knows a legacy of stewardship, and our kitchen scraps, the leaves of this autumn, and the embedded seeds of place are all an accumulation of the ages, left to tell stories that we can decipher. Our hands are the primary tools of

craft, and they can help us to trace and retrace the work of generations. Our taste buds also remember the savor, and the fragrance lingers like a balm in our nostrils.

There is a fine line between understanding the past, understanding ourselves, and cultural appropriation. It can be a tightrope walk. I approach my investigations as a curious gardener, eager to know tried-and-true precedents for using the plants I cultivate. As a garden historian, my hands love to retrace process as it had been carried out countless times before. I have grown to value the deep connection that comes when I save the first seeds of an heirloom harvest. I feel it when I take a soothing sip of the tea I was taught to make by my elders, when I replicate a 16th-century recipe for anise-seed cake, or when I process the mucilaginous root of comfrey (aka knitbone) for my annual batch of healing salve.

Researchers can now apply science and chemistry to regional foodways and folkways in order to better understand the wisdom behind the traditional uses of our botanicals. However, only a handful of the active constituents revealed in each plant can be replicated chemically, though many hundreds may be present. Having finally gained access to these studies (largely conducted in Germany and Japan, where pharmaceutical industries support transparency), we can begin to apply this knowledge when preparing traditional recipes, simple medicines, cosmetics, and fragrances.

As gardeners, we have the advantage of using fresh, whole plants for our food and medicine: bitter herbs for liver and kidney support, beets for blood, cranberries for vitamin C and urinary tract infections, and so on. Now the products offered by cottage industries are "value added"—our cocktail bitters are seen as beneficial, our borscht as a source of iron, our elderberry syrup as an antioxidant for seasonal well-being.

Each time I make my grandparents' pasta sauce with tomatoes, garlic, and basil I have grown, I bring the trade and immigration routes of the world, the chemical constituents of the plants, the soil of my backyard together in a simmering pot. A scent fills the house, just as it did when I was a child. Just as it did in my grandparents' youth, and theirs before them. We may not all be ethnobotanists, but the world is a richer place when we apply their principles to the plants, gardens, and communities around us.

FIDDLEHEADS

FIDDLEHEADS ARE THE EMERGING leaf frond and unfurling Fibonacci sequence of the ostrich fern (*Matteuccia struthiopteris*), rising above ground in defiance of winter as earthly art. They are among the earliest edibles of springtime, the Moxie of spring woodland greens, and a visual reminder that nature has long inspired our arts and crafts. They are also one of my favorite spring tonics. Fiddleheads have anti-inflammatory properties and twice the antioxidants as blueberries; they improve cholesterol levels and enhance immunity. Best of all, these pungent green fiddleheads taste of damp earth and moss, sunshine and spring.

When they first push through the woodland duff, they almost resemble hesitant noses, poking out of the cold ground to gauge safety before rising to meet the season ahead. Their curious beauty entices us to gather this seasonal side dish, literally following in the footsteps of generations of regional foragers before us.

Growing up, we would choose our steps carefully in the damp woodlands where fiddleheads grew, making our way around vernal pools, springs, and streams to select still-tight fuzzy fronds fit to take home. And we always made sure that we left enough behind to keep the fern colony strong. Walking back home, I would rub away the tannin-rich fuzz that discourages deer from eating them. Nowadays,

by the time they are on my kitchen counter, most of that fuzz is stuck to the burlap bag I will turn inside out to rinse them in. If you are not one for trekking through wetlands, ostrich fern is easily cultivated as a majestic shade plant for your garden, and if they approve of where you planted them, they will spread quickly and ensure that you have an abundant supply of fiddleheads right from your own backyard.

I wish I played fiddle, violin, or cello; I love the resonance and the sheer beauty of the instruments. But instead, I forage and cook the exquisite, emerald-green fiddleheads around me. If they are abundant, I pickle and enjoy them throughout the year. Most often I just harvest enough from my garden to make dinners special until fiddlehead season has passed. To prepare them, first quickly blanch them in boiling water. Then, on their own or in a stir fry, simply heat a little butter and sesame oil and toss them in a pan with some salt and pepper for a vernal side dish. I add them to frittatas and omelets, or to boiling pasta water, for the last few moments, to make a woodland pasta primavera.

Regardless of how they are cooked, they taste of the earth and greens overwintered in peatland ground. They even carry the savor of forest and a tannic tinge from the pine duff and sphagnum they emerge from. They are the asparagus of the woods, a luxurious native backdrop for any shade garden, and green garden art in the landscape.

FLORICULTURE

WHEN I was younger, I thought of floriculture as the weakest branch of the
botanical sciences. The grandmother of
horticulture. The doting aunt of agriculture. A sentimental Victorian arranging
flowers for her garden club. Botany was
then the only "appropriate" science for
genteel ladies. But fast forward a century:
women who held their spot in the work-
place after World War II were thrilled that

> Gardening is
> the art that uses
> flowers and
> plants as paint,
> and the soil and
> sky as canvas.
>
> —ELIZABETH
> MURRAY

canned peas were available to serve year-round. What was the point
of delivering homemade niceties to a family convinced that TV dinners were the bee's knees and that Birds Eye did it better? Commeri-
cal culture tried hard to create the image of the perfect wife, mother,
and consumer. Feminism and "having it all"—holding down a job,
raising kids, preparing meals, doing laundry—didn't leave a lot of
room for Aunt Flora to arrange flowers, make hollyhock dolls, and
overwinter dahlias. And video games and home computers further
distanced children and families from the natural world.

Happily, wise and passionate floricultural friends took up Flora's
mantle, planting seeds that helped create a generational shift. Women

like Sharon Lovejoy (whose books introduced nature crafts to a new generation), Jane Taylor (matriarch of 21st-century children's gardens), Alice Waters (who revived the school gardens movement), Betsy Williams and Tracy Kane (who did the same for fairy houses). All had a common goal: to get families out from behind blue screens and back into green landscapes, where they could connect in a frequently disconnected world and share experiences that are not for sale.

When I first began at Plimoth Plantation, few were teaching floral garden craft. Adelma Simmons and her Caprilands Herb Farm and some old-time garden club and herb society ladies seemed like vestiges, their lives centered around homes and gardens that they enhanced with seasonal garlands, floral potpourri, and lavender wands; they hosted high teas with floral butters, confections, and cordials that reminded us that life, like floriculture, has its sweetness too. It felt almost as if they had turned back the clock, to a time before world wars and economic depression.

When Martha Stewart arrived on the scene, many of my friends resented her deeply: they came from generations that knew the pressures and strains of underappreciated career women, striving to keep up at home. But some saw and valued a talented woman turning floriculture into a lifestyle brand. Over time, I came to credit Martha, Adelma, and all the quietly determined women who preceded them for preserving and reigniting connections to garden and environment that were in danger of becoming lost arts. In an age of consumerism, they reminded us that we enjoy a greater quality of a life when we live and craft from a garden—and that there is joy in the process, too.

Floriculture today means that our salves, salads, and cocktails are enriched with edible and medicinal flowers; that our participation in a "slow flower" movement is bringing back a locally grown florist trade; that we can fill our gardens with native wildflowers readily available from specialty nurseries and farmers markets. It means a child can learn to value his great-grandmother's iris more than out-of-season lilies from Chile or the Netherlands. And that an endangered pollinator will once again land on a native plant it had relied upon for millennia. Floriculture inspires an elementary school science class to visit their courtyard garden to study the parts of a flower and carry chive blossoms and rose petals back to the lunch room salad bar.

For me, it means that I don't just pinch off *Salvia officinalis* blossoms and toss them on the compost pile, I use them to make sage butter. I also make lavender soap, violet syrup, and floral ice cubes. I distill passing roses into rosewater, brew beer with the flowers of ale-hoof (*Glechoma hederacea*), and make vegetarian rennet for cheese with the flowers of lady's bedstraw (*Galium verum*) and skin ointment from *Calendula officinalis*.

A renewed appreciation for floriculture reflects the world's readiness for more feminine energy—witness artists like Whitney Krueger and Andy Goldsworthy, whose ephemeral creations would have been labeled flights of fancy in the patriarchal industrial age. If we pay attention, Flora's subtle charms—swags, nosegays, strewing herbs, flower petal mandalas—can entice us to play in the natural world. If we don't pay attention, we will miss a world of fragrance, beauty, and fleeting moments that we were born to enjoy.

These days, as I bask in my floriferous spring garden, sipping tea of clover blossom, mint, and bee balm flowers, I raise a cup in honor of Adelma Simmons, Martha Stewart, favorite aunties, groundbreaking friends . . . and the goddess Flora.

FORAGING

EACH FALL AT THE GARDEN CENTER where I worked as a teenager, a few rickety old vans would pull up at the edge of our parking lot. The elderly passengers who alighted, each with a basket or backpack, would quietly set off down an old logging road so nearly reclaimed by nature that I might not have known it was there if not for this curious crew. As I learned after several years, they were Portuguese immigrants; they would hike deep into the woods and, as the day's light was dwindling and we were closing the nursery, they would come away with baskets and bags brimming with mushrooms.

The annual visit of the Portuguese foragers held for me the same gravitas as the snapping turtles I saw returning to the exact same sandy spot in our field above the river, year after year. Hatchlings and annual migratory patterns. Enduring seasonal rites and rituals, grounding us in what to eat, where to find it, and the nature of home. Ancestral knowledge. Suddenly, I saw gathering mushrooms and dandelions with my grandmother, grapes and walnuts with my grandfather, with different eyes.

Years later, when I was traveling with friends in Italy, we saw people pulled over along mountain roadsides gathering fall fruits, chestnuts, truffles, and herbs. Nearby, where the mountains became hills, generations of household lands, crafted to leave an enduring legacy, sloped down to the sea. Some were designated exclusively for orchards and vineyards, but there were also gardens of useful plants, grown from the knowledge of what the land and seasons could support, within reach of every kitchen. Drifts of chamomile and fennel, cascading rosemary, little groves of olive, fig, hazelnut, and bergamot, prodigious shrubs of lemon verbena, quince, and bay. If there was a space to plant, it was enriched with edible beauty, used to soothe and nourish through countless seasons. My eyes filled with tears; I could clearly see how the landscapes and skill-sharing work of ancestral foragers had informed the diet and traditions I had inherited from my grandparents.

Don't believe everything you read in the papers: foraging is not the exclusive domain of Brooklyn hipsters. In reality we all grew up foraging. Strawberries in spring, blackberries and blueberries in summer, autumn fruits and nuts. Maybe even nibbling on wintergreen in the coldest months, when it seemed nothing was alive. When I forage, I feel I am rediscovering forgotten places and gifts left by the people who lived here before me. Establishing a connection to place as deep as the native plants I encounter, and as farflung as the immigrants who settled and brought the plants they loved to the landscape before me. Generations of recipes, customs, and seasonal foodways.

I try to forage with reverence and respect. I imagine who might have first planted the rhubarb or pear tree, or gauge how much of the abandoned horseradish I can gather while leaving enough for future harvests. I contemplate how an entire field came to be seeded in valerian, and wonder if the surprising woodland stand of daylilies and Solomon's seal is a relic dooryard garden. Gathering walnuts, I think of the people who had done the same before me—did they use them for protein, dye, cookies, or an old-fashioned wedding cake?

Foraging can mean returning to the same stand of berries we knew as a child. Or, now that we're all grown up, using chokecherry for a cocktail, staghorn sumac for a spice blend, black trumpet mushrooms for an omelet, and groundnuts for a winter stew. A cup of wintergreen tea to refresh the spirit and help aching knees.

If we eat from the land, we tend to care more deeply about environmental stewardship and the habitats we harvest from. And certainly, more than most supermarket shoppers, people who forage are attuned to the ebb and flow of seasons and the natural environment. Nevertheless, poverty has led to overharvesting by some foragers, as has harvesting for market (restaurants, breweries, grocers, pharmaceuticals); such pressures can do great harm to regional habitats. Better to follow the wise First Nations perspective on foraging: pass the first plant you find and keep searching until you find a place where

the plant grows in abundance. Then carefully glean only what you need, leaving a resilient patch behind. Foraging is best applied in places with endlessly abundant plants like clover blossom, dandelion, nettles, violets, and alehoof, or abandoned orchards. Even better, in the process of harvesting knotweed, garlic mustard, and bishop's weed, we are eating healthful plants while helping to eradicate problematic invasives from wild places. Do not, however, forage these (or any plants) in areas that might have been sprayed with toxic chemicals.

If by chance you yearn to harvest more than your environs can supply, you have identified the perfect opportunity to grow another local industry. In these days of farmers markets, your own backyard garden may just offer an appropriate habitat to grow native edible and medicinal forage. When we plant our backyards and small farms with elderberry, fiddleheads, ramps, and wintergreen, we are not only sparing our local habitat from overharvest (and providing for our tables and wallets)—we may even be repatriating some old botanical friends.

GARDEN CRAFT

> To forget how to dig the earth and tend the soil is to forget ourselves.
>
> —MAHATMA GANDHI

IN just a few generations, most Americans lost daily connections to the land. In the process, many of us lost our ability to ease and nourish mind, body, and spirit from the landscape. Instead, our elemental connections to seeds, soil, and sweat were traded for the allure of convenient consumer goods. Once our homesteads became nonessential contributors to our lives and livelihoods, we yielded to foundation plantings and lawns. These new suburban landscapes require little more knowledge than how to turn on a sprinkler or plant a window box. Now we hire crews to mow and blow, with hopes that the chemicals that they use won't be too toxic, and that the droning leaf blowers arrive when we aren't at home.

As land-based life skills slowly slipped away, we became disconnected from the reciprocal relationship that exists when we lived in concert with tides, seasons, and soil. For a time, post-war consumer culture lulled us into believing that homogeneity tasted better than homegrown and seduced us into thinking that rapid delivery was more gratifying than sweat equity. Fortunately, there is a growing awareness that this kind of detachment comes at a cost to the environment and the human spirit. I'm quite sure that the majority of our population wouldn't be content farming, but most of us connected with the land don't mind sweat, or dirt under our fingernails, if we are able to trade in our long commutes, microwaved food eaten on the run, and a fluorescent-lit office cubicle. We would rather raise kids with a sense of purpose and a connection to land and sky than a generation waiting behind a blue screen for the zombie apocalypse.

Before DIY workshops and how-to videos, we planted seeds and worked with mentors in our family and community. I am grateful to see a renaissance of new college curriculums and internships, online master gardener programs, and makerspaces that offer skill sharing, but the need to rekindle our daily connections to garden craft remains. Too many of us long to reconnect our hands with meaningful work, the warp and weft that bind and connect us to the past. Over and under. Seed to table. Forage to fire. The flavor of seasonal recipes, heirloom produce, and terroir.

Malls are dying, and local markets are back at the core of thriving local communities. New generations that care more about sustainability and integrity are taking back the reins, and many men and women who were restructured out of well-paying jobs or replaced by automation have found fulfilling work in their local economies and are putting their money where their mouth is by making every purchase possible count. These gig-economy artisans craft livelihoods from the landscapes around them, while demonstrating meaningful ways to keep hands busy through the seasons, feed our families, and ornament our lives. They are local farmers and gardeners—craftspeople who know which wood makes the best home or boat, flute or fencepost, salad bowl or mortar and pestle. Not least, these new economies foster quality of life improvements, including healthier

air, increased food security, and more vibrant communities that find nourishment in a sense of place.

Everything has its season, but I trust sustainability's future because it meets bottom-line goals. Until we rebuild a middle class, new generations will need to know thrift and ingenuity. Craftsmanship that pays well can substitute for what we came to identify as industry—a new age of human industry that completes the shift from an inhumane industrial age to land-based craftsmanship and cottage industry. Jobs that come with mastery and respect. Jobs you can juggle with raising a family while doing work that feels meaningful.

There is something for everybody in the new marketplace for garden craft; and its cottage industries, besides clearing paths to a more sustainable future, are helping us to preserve heirloom crops and revive artisanal crafts that industrial agriculture would have allowed to die. Trades like beekeeper, greengrocer, florist, basket maker, broom maker, orchardist, herbalist, apothecary, perfumer, confectioner, brewer—all can produce a wide range of everyday, holiday, and seasonal market items. Craft products include goods like artisanal pickles and preserves, botanical pharmaceuticals and beeswax candles, balms and bitters, compost and cosmetics, propolis and perries, to name just a few.

Craft doesn't come easily to everyone, but if we are fortunate, a skill that we have a passion for can become a source of income, or even a career. It certainly has in my case. Instead of purchasing cheap aluminum tools for my work, I restore antique or quality tools designed to last. In fact, I enjoy using power tools about as much as I do vacuuming; they make me feel more powerful than I should, and I often find that they do as much injury as good in the landscape. The less I use them, the more I enjoy real and collaborative connections with soil and grain, handwork and habitat. I also don't feel bound to purchase plastic garden furniture, arbors, or fencing; I know how to make these things for free from the surrounding landscape—and I've learned to ignore the advertisements that tell me I don't have time for it. My greatest pleasure comes from working in my own landscape, saving drive and shopping time, petroleum, and cost to me and the environment, by simply making my own (or buying from artisan friends). The result is a garden that reflects my own skills, materials, and taste.

We can still craft a sustainable future before we have lost our local farms, or filled our landfills, oceans, and bodies with plastic because we were told it was easier that way. Craft enables us to reclaim and revel in the process, but we all live modern lives, and we need to pick and choose the crafts that bring joy. Wisely, younger generations of digital natives are beginning to see the internet as just another tool in the artisans' kit, and sustainability sciences as industries. So, let's blend old and new tools together and remember the joy in the process of rebuilding the common ground we share. Remember what every kid in a school garden knows: there is nothing sweeter than the fruits of your own labors. Particularly when they are a reflection of your own garden craft. Garden craft and artisanal agriculture *are* the new American industry, and they are providing handwork made to nourish and to last.

GRUIT

BEFORE THERE WAS BEER, the English drank ale and Germans gulped gruit. Both beverages, ale and gruit, were crafted like beer at a time when hops was not in common use. Instead they were made with traditional seasonal botanicals—alehoof, mugwort, bog rosemary, wormwood, sage, spruce, hawthorn, elder, and other herbs and fruits. But since hops increased shelf life and stability, it slowly replaced earlier traditional methods. In the process, we lost many seasonal and regional botanicals, and a tremendous range of flavors. Until recently...

Some time ago, at a folk crafts festival in Germany, I realized that beer could be many things, including really delicious. At Plimoth Plantation, in the name of special research and interdisciplinary experimentation, we malted, brewed, and drank traditional ale and period beer made from the heirloom hops and grains we grew. At Strawbery Banke, we partnered with Portsmouth Brewery to make a beer using green hops freshly harvested from the Colonial Revival garden. Later, I worked with several other breweries, including Earth Eagle and Dogfish Head, experimenting with heirloom produce and foraged herbs, and even recreating New England's first documented beer. Every project sated a growing thirst for a taste of place.

Smack dab in the middle of our local agriculture and craft movement is local beer and gruit. Breweries grew up alongside the

cultural shift away from drinking and driving; and brew pubs put the neighborhood back in neighborhood bar, making it a lot easier to walk home instead of drive. The wake of this overnight success sent out ripples to local farmers, who once again had markets for heirloom grains, hops, and botanicals. Spent malt from the brewing process circled back to farms to feed the hogs that made the brats and sausages served as accompaniments to our pint. The same ripples engaged and trained up new generations of foragers and gardeners to harvest and provide traditional herbs, fruits, and seaweed needed by brewer and mixologist alike. And suddenly we can all raise a glass to the quality of life our new local economies are helping us to enjoy.

Samuel Adams might have been the only American craft brew when it was launched in 1984, but craft beers now make up more than twenty-five percent of the overall U.S. beer market. And they have helped virtually every community across the nation gain access to all that local can mean. Cheers to the years!

HEIRLOOMS

WHAT PLANT MADE YOU FALL in love with gardening? Often, it is one that was handed down to us—a grandmother's morning glory, a beautifully mottled Native American bean of great antiquity. Perhaps an inherited rhubarb patch that came with your house. Most of us associate family heirlooms with dishware, jewelry, or old furniture valued as sentimental antiques; but heirloom plants are our *cultural* inheritance, from a time before chemical agriculture, a time with a much greater diversity of flavors, a time when our food was grown and nurtured with love. Heirloom plants may show less uniformity, but that funky-looking plant comes with a resilient history of adaptation, perseverance, and resistance; a recipe box of flavor; and a treasure trove of stories that preserve history and keep healthy alternatives alive right in our own backyards.

In contrast, most food plants selected for agribusiness are chosen for uniformity, durability, and long-term storage. Flavor, texture, and fragrance are secondary considerations. Primary to big ag is that a hybrid tomato can be mechanically harvested unripe, endure being shipped thousands of miles, and still arrive perfectly round. And we wonder why they go from stonehard pink orbs to mush overnight.

Remember the flavor of sun-ripened heirloom tomatoes, warm from the vine? Can you taste the sweet diversity of flavors from the earliest cherry tomatoes to the largest oxheart or Brandywine your region could grow? That is because in addition to fostering biodiversity, heirloom plants preserve taste memory and forgotten foods worthy of preservation. When we save these seeds, we are participants in a long line of stewardship. Their fruits are truly gifts from the past, and they are always sweeter for it. My grandfather grew his own tomatoes and other veg; like every generation before him, he took special care to save seeds from the earliest harvested, or from those that came from a plant with a higher yield or disease resistance. He was selecting for the best traits, adapting the crops he grew to the needs of his own backyard and kitchen. If he ate something he particularly liked, he would express his delight and smear some of the seeds on a brown paper bag to dry. When they did, he would tie up each precious bundle with string, label the bag, and save it to complete the cycle, planting that runner bean or pickling cucumber again the next season.

Unlike hybrids, heirloom plants are not genetic clones but expressions of the place-based values of previous generations. They provide living links to the success stories behind crops that people like my grandfather or the farmer down the street created in the process of improving and preserving genetic diversity, flavor, and taste. Until recently, seed saving from open-pollinated plants was a practice carried out annually for thousands of years, in cultures across the world. Not the seeds of clones, but careful hand selection in a near sacramental process that preserved the most bountiful and enduring varieties with every passing season: perhaps a preference for the tall blue corn, the most prolific cucumber, the first melon to ripen, the tomato that didn't split, the squash that overwintered well, the fragrant heliotrope or savory cilantro. Each region had its favorites.

After the world wars, chemical companies began to buy up all our regional seed houses. They focused on seed selections adapted to the needs of agribusiness centered in the Midwest—and patenting hybrids that could not be saved by farmers or home gardeners. But home gardeners continued to prefer open-pollinated diversity, which yielded staggered harvest times, shape, color, and ultimately regional success; we generally want our spinach and radishes to come in over the course of weeks, whereas the mechanized harvesting, shipping, processing, and marketing practices of agribusiness all favor uniformity over flavor and diversity. Unsurprisingly, small seed companies without tremendous marketing budgets and international distribution rapidly disappeared. "New and improved," flash-in-the-pan, replacement hybrids kept arriving on the scene, and thousands of years of seed-saving wisdom and regional diversity disappeared. So too did our cultural inheritance, handed down freely and adapted to place through previous generations.

Hybrids are genetic clones; their seeds can't be saved and grown again. As clones, an inherent danger is that if one plant is blighted, the entire crop is lost. On the other hand, biodiverse heirloom plants ensure surviving plants and enduring seeds that provide backup for a more resilient future.

Taken a step further, GMOs have shifted the conversation to one of ethical choices and food sovereignty. Many GM crops are developed to eliminate the need for farm workers, to foster a reliance on chemicals sold by the same companies, and to create foods that maximize profit. All around the world, gardeners are fighting for the right to save their own seeds and traditional food crops. We don't want commercially manipulated and processed seeds, any more than we want the processed foods that they invest so heavily in force-feeding us. Perhaps better than anyone, gardeners understand that food is

more than just calories, and that
the things we grow and ingest
inform and communicate with
our DNA and tell it what to do.

In a 2012 interview with Bill
Moyers, the brilliant Vandana
Shiva said, "You cannot insert
a gene you took from a bacteria
into a seed and call it life. You have not created life, instead you have
only polluted it." But ever the optimist, she goes on to say of heirloom seeds, "The garden teaches us that there is something we are
all capable of doing. Only with something so small that can be held
in everyone's hand can we challenge the empire."

Traditional heirloom plants and vibrant local agricultural economies are the keys to a counterresistance. Unless cross-contaminated
by GMOs, open-pollinated heirlooms enable each of us to keep food
safety and food sovereignty alive. They also help make our communities resilient in the face of the next potato famine, corn blight, or
oppressive regime. They offer sanity in a seed and unleash a gardener's superpower to outgrow unfettered capitalism in backyards all
around the world.

Planting a garden with heirloom seeds is not only an act of preservation, it's a living connection to those that have come before us,
and a brighter future for those who will inherit what we keep alive.
If carefully tended, these seeds will be a part of our future. Annually, our gardens remind us that with a tiny seed, we can preserve a
democracy, we can engage a child, we can nourish a family; we can
protect our environment, challenge a flawed system, preserve our
cultural inheritance—and savor the difference.

And hope for the future can always be found where we plant seeds.

HERBALISM

BEING AN HERBALIST does not have to be an occupation. If you are simply a person who collects and grows herbs, or a person who practices household wellness by using herbs from your own garden, you probably don't—and probably shouldn't—prescribe. But there are basic principles, systems, and safety measures that herbalists need to share in order to demystify the craft.

You may already be a budding herbalist if you explore the herbal world with the curious eye of a gardener, cook, or parent. These days, chefs, mixologists, brewers, cosmeticians, and CBD purveyors are all opening gates to the herb garden and helping our culture rediscover many of the essential flavors and health-giving properties of locally grown herbs. Since many of our traditional botanicals have a history of culinary use, gardeners and cooks already have a leg up: we are familiar with the cultivation, flavor profiles, and benefits of herbs. And now we have science as well as lore to help us make informed choices.

Herbalism is born of gardens. It blends horticulture, foodways, folkways, art, science, and craft for the sake of wellness. It is the gardener's first line of defense, and, while it can sound mysterious, it can be as simple as chamomile tea for digestion, jewelweed for poison ivy, or aloe for a burn. It is the craft of harvesting plants with useful chemical components and turning these plants into helpful medicine.

In traditional communities, an old sage is a wise person who knows how to heal mind, body, and spirit, often with herbs. Unfortunately, ours are among the first generations that have a significant gap in herbal knowledge passed down from elders. When government regulatory agencies took over medical oversight in the early 20th century, there was an ugly divorce between the pharmaceutical industry and herbalism. Without question, science had brought new standards to medicine, but just as notably, herbs could not be patented for profit, and the pharmaceutical industry embraced the market shift to promote a distrust of herbal medicines and the people who dispensed them. Instead, a flood of chemical medicines from an emerging pharmaceutical industry virtually swept away the libraries and sage elders that carried the wisdom of the ages. Meanwhile other nations around the world, far less willing to throw the baby (or elder) out with the bathwater, kept traditional herbal preparations on pharmacy shelves alongside chemical medicines.

Fortunately, with a heavy lean on the modern chemical analysis of herbs, we are reconstructing old systems of understanding that can be partnered with current scientific method to create a more complete picture of medicine from the garden. A rising generation of science-literate herbalists and nutritionists are offering plant-based alternatives, and we are once again recognizing plants for both their nutritional and medicinal value and integrating them into medicine cabinets, cosmetic counters, family meals, and cocktail bars.

Modern herbalists need to know which plants can be sustainably gleaned or foraged; as entrepreneurial herbalists, we must have the wisdom to foster an environment where we can cultivate more than we harvest. If we know that we need more of a botanical than we can harvest sustainably, we will either plant more, cultivating the plant ourselves, or hire a local farm to raise a crop for production that helps us, the farmer, and the land to thrive.

Once we have access to plants, herbalism becomes an engaging exploration of craft and science as we ferment, distill, and blend the salves, elixirs, and fragrances that serve as our primary on-ramps to wellness. Herbalism means that my teapot is a seasonal prescription for almost anything that ails me—or at least a steaming cup of calm to help me move through it. Second to the teapot, my salad bowl offers medicine fresh from the garden, accompanied by the roughage we all need to keep healthy; it is a blend of seasonal flavors and medicinal attributes delivered up as a tasty mix of greens, veg, roots, fruits, flowers, seeds. The rest is just a matter of applying herbalism to the dishes I cook. Savory to counterbalance "windy" foods, thyme and garlic as natural antimicrobials, and so on.

Long ago I learned that I seldom needed maximum-strength over-the-counter drugs to blow out a sniffle or calm a headache if I gave my body the chance to respond to the subtlety of herbs that I now routinely rely upon to boost my immune system and act as my first line of defense. Herbalism reminds me that aromatherapy starts with smelling a rose to gladden the heart, planting a lilac to welcome spring. Herbalism can be as simple as making a pot of elderberry-sage tea, or as complex as distilling or compounding medicines for market. But at its heart, herbalism means applying the science and craft of botanical wellness, right from our own garden.

INDIGENOUS PLANTS

IN RECENT YEARS we have begun to rekindle and deepen our interest in native plants. This represents a healthy movement to cultivate flora that provides traditional food, shelter, and medicine for our native fauna—heirloom gardening and culinary tradition for the birds and the bees.

In many parts of the world, nativity is difficult to define; it's easy to overlook many pieces of a layered and complex history of travel, settlement, and trade. For instance, it would be hard to imagine a plant more ubiquitous than *Rosa rugosa* on Cape Cod. It thrives along the sandy shores and dunes and is now an emblematic staple of seaside gardens from Provincetown to Plymouth. So I was stunned to find that pollen and seed analysis dated the introduction of *R. rugosa* to the 13th century—an impressive pedigree for a Cape family but not one that ran as deep as the seldom-seen natives, *R. carolina* and *R. virginiana*. It challenged my notions of what native really means, and while I now understand that it had centuries to naturalize, it made me less stringent about who did and didn't belong in the tribe.

And native echinaceas might have been a familiar sight in the Midwest, but they are as much a foreigner to New England as I am.

A New England aster or cranberry that comes from wild-collected seed stock and soils of this particular spit of land carries more weight than a prairie coneflower (*Echinacea purpurea*) because they hold the genetic makeup and stories of this region, where they have been adapting for centuries. For me, more than native, the word indigenous speaks to the endemic plants that help define regional habitat.

That being said, the Three Sisters agriculture of my region originated in the Native peoples of the Southwest. First beans, then corn and squash, defining a shift from hunter-gatherer to agricultural people for well over a thousand years. These agricultural siblings represent seeds, plants, and cultural practices that were adapted by tribes as they followed along trade routes and new settlements patterns, all the way to New England and beyond.

I don't know that there is a right answer, but I do value the cultural awareness that goes along with our latest native plant movement. We often take for granted the things of our daily life. The same roadside New England aster we looked right past until recently was lauded in England as the elegant Michaelmas daisy by 1640. Similarly, our goldenrod arrived in 17th-century European gardens and apothecaries as an exotic flower and medicinal plant. After centuries of populations devaluing and eradicating plants like milkweed, we have learned the hard way (from the still small voices of monarchs, pollinators, and perfumers) that we don't need to, and often can't, buy the best things in life. Sometimes we just need to make sure that we don't mow them down.

So, is the dandelion, introduced by early colonial gardeners, any less important to native bees searching out food in the early spring? We have all met earnest native-plant Nazis eager to tell us why we *must* grow only natives. I will never be that much of a purist about anything. Instead, I see place-based knowledge as a critical way to touch base with, and live peaceably in, our environment. Whether learning from indigenous peoples, the native plants, or the soil sciences we share, we are all relearning how to read the landscape and remember the old names. Sassafras, wuttahimneash, saskatoon. These are times for learning sustainable ways to live in concert with the plants and animals, the seasons, the breezes, the ebb and the flow. What is most important is sharing the outgrowths of plants that define a region and rebuilding a web that helps to feed and shelter an ecoregion.

Our present building and landscaping boom presents an opportunity to preserve and enhance the spaces we have left with all the plants in the web. The wee ones along the path and the ancient ones with canopies among the clouds. Planting indigenous recreates vibrant ecosystems and communities. It respects a heritage not erased but made precious and sacred by scarcity. Ice plant and cedar. Salamander and walking stick. Sacred landscapes reemerging from fallow ground to preserve homeland, habit, and diversity.

Since the difference between value and profit is often lost in late-stage capitalism, box store sales will likely never embrace true natives, which too often fly in the face of industry conventions. That means growing an industry for place-based natives falls to you and me, and the few remaining fine garden centers. Seeding jack-in-the pulpit into our backyard, cultivating a line of native plants for your local farmers market or garden club plant sale, designing a wildflower meadow, carefully transplanting and saving the clumps of lady's slipper otherwise doomed by the developer's bulldozer, or changing legislation that replants wildflowers instead of mowing our endless highways.

And have faith. When we reintroduce natives, we are helping them to spread far and wide, in the ground, on the wind, and maybe, like the rugosa rose, in ways we can never even begin to imagine.

JACK-O-LANTERNS

A LANTERN CAN SHED LIGHT across many cultures and time periods, but few grin back across the generations like a jack-o-lantern. The earliest lanterns (lanthorns, lamp-horns) were made from nearly translucent sheets of horn, and later, vegetables. What began as the representation of a glowing spirit on All Hallows' Eve didn't have its roots in a pumpkin patch. The first jack-o-lanterns were turnips, carved with grimacing faces and set out to protect homes, pathways, and ceremonial fires.

Long before immigrating to the New World, Celts and English revelers kept All Hallows' Eve (the earlier Celtic Samhain) with jack-o-lanterns carved from large root crops and gourds. On Samhain, the veil between the human world and the spirit world was said to be lifted. Ghoulish carved faces that shone light (from glowing embers or candles) through a grimace were intended to scare evil spirits back into submission as they tried to walk the earth again—among them the infamous Stingy Jack, a farmer who, having struck a deal with the devil, was doomed to walk the earth forever. Another Jack, French explorer Jacques Cartier, sailed to present-day Quebec via the St. Lawrence River, where he reported finding *gros pompions* ("large melons") in 1534. Over time in the New World, larger late-harvest pumpkins replaced the Old World roots and gourds.

Whether you grow your own jack-o-lantern pumpkins or turnips, Halloween traditions offer a fun way to engage kids in the act of growing plants, enjoying craft, and eating fruit and veg. Some of the oldest heirloom pumpkins include the Long Island or Magdalena Big Cheese (because the shape was thought to resemble a wheel of cheese), New England Pie, Connecticut Field, and Long Pie; several other heirloom pumpkins evolved for hundreds of years in Europe before returning to the states in much altered form, color, and shape.

You can buy heirloom pumpkin seeds to grow your own, or if you find open-pollinated varieties in your local farm stand, you can carve them and save seeds to plant the following spring. Of course, it's always good to give into temptation and roast some of the seeds to enjoy as a fortifying snack: these pepitas, as they are also known, may help absorb all the sugar that follows on the heels of trick-or-treating.

In many parts of North America, pumpkins are virtually the poster child of agritourism. Farm grown, pick your own, carving contests, chucking contests—there are even pumpkin regattas, where people turn enormous hollowed-out pumpkins into boats that they "race." Once the festivities are over, you don't have to leave your jack-o-lanterns to rot on the doorstep. A fun family activity to work off the sugar rush is to plan a somewhat more wholesome baking project together. It's a delicious way to teach kids valuable lessons about food waste while making your own pumpkin bread, butter, soup, or pies. One of the earliest recipes for pumpkin is for a pie, in which the "pompion" is sliced up like an apple and spiced with ginger and butter (and likely a little sugar) before closing it up in a "coffin" of pastry. Once apples were introduced to America, so too was the brilliant method of filling the emptied cavity of a pumpkin with apples and then baking them all together, much as I do when serving soup in pumpkins today. I can hardly think of a better way to keep evil spirits at bay.

JOHNNY-JUMP-UPS

MY CHILDHOOD ENTRÉE INTO BOTANY went something like this: *Viola tricolor* is a viola with three colors—as simple as that. I loved the name and the way they burst forth defiantly to show their bright purple, yellow, and white faces amid gray landscapes. And I could eat them too! Johnny eating flowers. I loved that! Even in the northern climate where I was raised, these merry harbingers of spring would self-sow and begin to bloom so early that they would survive intermittent snows. Somehow eating flowers that could defy snow seemed even cooler. I had a companion plant; and that's foundational stuff for a kid—in life, in growth, and in caring enough to cultivate better things as I grew. I soon learned other scientific names, and sure enough, the binomial system was less complex than I might have imagined. *Viola odorata* was the name for—you guessed it— fragrant violets. Quickly, this budding horticulturist came to see how easy it was to have a basic understanding of taxonomy.

An early common name for *Viola tricolor* was heartsease. According to colonial-era herbals, its three petals were likened to the valves of the heart, which likeness (according to the old doctrine of signatures) was a sign that heartsease was beneficial to the heart. Today, we prescribe pharmaceuticals, we drink, smoke, and disappear for endless hours behind blue screens to gladden or numb our hearts.

As a gardener, and a believer in the Hippocratic oath, I would rather begin with the simple aromatherapy of stooping to inhale a flower's intoxicating scent, followed up by a salad of medicinal greens and edible blooms. Does it actually ease and gladden my heart? Of course. Does it work as well as a prescription? I'm grateful to say that I don't know. I just know that heartsease comes in many forms and flavors from my garden, and they all make me smile.

These days, the humble johnny-jump-up wins my favor because it knows how to integrate itself into my perennial borders, vegetable gardens, and kitchen. The older I get, the less I care for fussy flowers and botanical divas, and the more I care for the honest heirloom and nimble native. Plants that serve my household, wildlife, and pollinators alike. Plants like johnny-jump-ups that do not require tilling, hybridization, or even annual replanting. I have come to favor plants that are reminders of the stories, flavors, and remedies that can be shared between gardeners, cultures, and centuries. Any plant that brings my heart ease.

KELP

MANY PEOPLE'S FIRST ENCOUNTER with kelp likely came while beachcombing or swimming in the ocean; but for coastal dwellers, it may well have been in the garden. In New England, old-time gardeners mulched vegetables and flowers with seaweed; if storms threw an excess upon the shore, they added it to their compost.

Seaweed is teeming with life. You can see it and smell it. It brings the vibrant outgrowth of the sea to our gardens. Unlike most fertilizers it can be added directly to the soil, where its nutrients are readily bioavailable to plants. Seaweed brings enriching nutrients to tiny microbes, fat earthworms, extensive root systems, and huge cabbage. Aquatic life meets earth in a garden bed. They kiss and make the most beautiful babies, angelica to zucchini.

The bioavailable nutrients in kelp include nitrogen, phosphorus, and magnesium. Nitrogen is used by plants for producing leaf growth and lush green leaves. Phosphorus helps plants to develop strong, deep root systems and increases fruit yield. Potassium improves flower color and size, as well as a plant's overall strength and resilience. Kelp offers the same primary ingredients chemically replicated in commercial fertilizers but without any of the questionable chemical ingredients and runoff attributed to commercial regimes. Instead of a chemical soup, seaweed sets about photosynthesizing from surface sunlight and, in the process, grows a gelatinous green fertilizer that our gardens love to absorb. Like a healthy diet for humans, the nutrients in kelp are essential for growth and keeping up defenses.

Kelp also brings more than sixty trace minerals to the soil, including iodine. Iodine aids in biomass production and increases the antioxidant levels in plants, which makes them more resistant to drought and stress. Other trace minerals promote microbial activity in healthy soil. Compared to some of the nearly sterile soils present in many agribusiness fields, rich organic earth (especially when enriched with organic matter like seaweed) holds more life in one teaspoon than we could ever count or define. But that's the cool thing about science: discovery is endless, and we will continue to learn forever. We keep finding more common language and, if we are fortunate, more common ground. Using the seaweed principle, we are like that teaspoon of soil, teeming with life. Seaweed is sloppy democracy. It requires individual participation, and everyone working at the microbial level to enrich common ground. By using thoughtful organic methods, we foster healthy, vibrant growth for all life around us.

Many chemical regimes render soil nearly sterile, but shock plant populations to life with quick doses of junk food—junk that can leave numerous biohazards in its wake. In recent years, we learned that chemical fertilizers were created to use up and profit from weapons waste after World War II. There is wisdom in knowing how to dispose of our waste, but using war chemicals on the ground is toxic for plants, humans, and the earth. These synthetic fertilizers have no proper application on land that supplies food.

And yet industrial agriculture is force-feeding us chemicals in the name of affordable food, even though science shows that many

of the chemically based fertilizers and pesticides carry hidden health costs and environmental consequences. Many organic gardeners point out that the salts that make up the bulk weight and water solubility of these granular fertilizers also kill beneficial bacteria and fungi, and even sterilize soil over time. These synthetic fertilizers are not bioabsorbed; they eventually make their way into rivers, lakes, and oceans, causing algae blooms that deplete oxygen and leave vast dead zones in their wake.

In the process of rethinking systems, it becomes critical that we explore historical precedent and proven best practices to heal ourselves and the earth. Seaweed is a familiar and effective remedy to some major cultural woes. Besides fostering deep root development and prolific leafy veg, a dense mat of dried seaweed can serve as a mulch to deter weeds—all benefits that help to prevent erosion and runoff that challenge planetary well-being. And sustainability requires that we further explore managed harvests that are mutually beneficial: kelp and other algal forms can be grown and harvested as renewable resources that benefit the health of humans and the earth.

And a new twist: there are sheep on the Orkney Islands that, for lack of pasture, have evolved to eat seaweed along the rocky coasts. As a result, the concept of feeding seaweed to cows has gained traction in recent years; studies have demonstrated its potential to cut back on methane. About a quarter of the methane in the country comes from cattle when they belch or fart. Knowing of these adaptations among sheep, people along New England's coast are exploring ways of blending seaweed into organic feed as a means to reduce global warming. And as they say, what is good for the goose is good for the gander.

Seaweed, like most herbs and vegetables, offers good medicine and sound nutrition for our bodies. We may not think we've eaten much seaweed, but it's an ingredient in lots of foods we love: candies, pudding, jellies, bakery and dairy products (my County Clare grandmother used Spanish moss, a kind of seaweed, to make ice cream). It's also added to processed meats and sausages, used to clarify wine and beer, and is even found in toothpaste, shampoo, and many pharmaceuticals. And in these days of dwindling fish stocks, sea veg offers a healthy and more sustainable alternative to our floundering fisheries and coastal economies; I even know of someone who paid for her college education harvesting seaweed!

Many of the most nutritious plants, disparagingly called weeds, carry terroir, the flavors of the land and the season. Well, kelp, like oysters, has merroir—think of it as umami of the sea. And it is there for any coastal forager. Like other vegetables, sea vegetables come in a wide variety of flavors and textures. Some are akin to collards, others more like green salt. Seaweed can be a complement or an accompaniment; it is seldom a main course. Kelp is like gelatinous kale, and it is perhaps best when sliced thinly. Like meat, it adds a depth of flavor and density to soup or broth as well as a subtle element of the briny ocean. I love using it instead of salt.

In the 19th century, the Frank Jones Brewery purchased 120 tons of seaweed annually for brewing beer; in the same town today, the Portsmouth Brewery makes a similar beer using the same kind of marine alga. Sugar kelp can add wonderful flavor, though its primary use in beer is to help clarify what would naturally be a cloudier brew. Like salt to caramel, it adds an unexpected layer, a wee bit of saltiness that makes me want to drink a little more—maybe even raise a pint to seaweed and all it can do to buoy our fisheries, enrich our seaside landscapes, and grow a more sustainable future.

KITCHEN CRAFT

The cooke is half a physician.

—ANDREW BOORDE

KITCHENS tend to be the primary gathering place in most homes, and for the gardener it often serves as our horticultural experimentation center. Our playroom, equipped with everything we need to process the seeds, flowers, vegetables, and fruits from our landscape. Kitchen craft is a natural outgrowth of garden craft. When the harvest is abundant, crafty gardeners know how to put it up in jars and freezers for the winter to come.

Countless dollars have gone into convincing us that cooking and preparing food is too much work and better left to the industrial foods industry. Over and over, Kentucky farmer-poet Wendell Berry's writings remind us that eating is an "agricultural act"; unfortunately, industrial food factories employ more chemists than growers or chefs. Chemists are welcome at my table, but I have no interest in them creating my food. As a gardener, I want that role to be mine.

Kitchen craft is knowing the process of cooking and why it can be simpler and more affordable to prepare real food than it is to drive to the store. As a child of parents and grandparents that lived through world wars and the Great Depression, I was always acutely

aware of how they survived. They knew how to supplement their diets with what they grew, and they knew thrift.

When the coronavirus hit, some consumers were so out of touch that they were hoarding toilet paper and frozen pizza and lamenting restaurant closings—instead of buying sacks of whole grains, produce with shelf life, and seeds. Fortunately, many families did eventually set about the work of planting seeds to ensure food security. Others took time to teach their kids how to cook meals and bake for the first time. In the process, they turned cooking into science, math, economics, and social studies projects. They held class outdoors and addressed STEM requirements from the garden. Other friends took to reading labels and turned lunchtime into a nutrition class. Reading labels can do a lot to inspire home cooking; and lessons in nutrition can send almost any consumer running back to the garden as often as possible for wellness.

The idea that food is medicine, that you are what you eat, has served us well throughout most of history, and if I were the betting type, I would put my money on a homemade quiche with fresh garden herbs and veggies over boxed mac and cheese or a frozen dinner any day.

There is a craft to cooking whole foods. Like any other skill we know by rote, there is also a meditative side to the gathering, chopping, stirring, and seasoning that go into an artful meal. Maybe not every day, maybe not picture-perfect ingredients, but wholesome meals. Pots and pans instead of boxes and cans. Cook in or cookout instead of takeout. Knife skills for a generation that has known only plastic forks and a world of fast food. A family learning together that the soil on your radishes is the healthy alternative to sprayed produce that has traveled thousands of miles in the hands of numerous middlemen.

As we learn more about all that is wrong with processed foods, more and more of us are embracing the joys of home cooking. New generations, not relegated to kitchen slaves or requisite housekeepers—but creative cooks. Men, women, and children working together to nourish family. Kids whose school garden programs connected garden to kitchen, kids now excited to eat the things grown together or bought from a familiar farm.

As gardeners, we may grow a bumper crop of beets, but that doesn't mean that we suddenly become insta-chefs. But when we are comfortable in our kitchens, we know that a glut of beets can be pickled, juiced, or frozen, or turned to borscht or a simple salad. We used to call that home economy. Kitchen craft can be as simple as preparing one thing from our garden each day. Simmering local beans to serve with stuffed peppers on Saturday. Sunday supper with veggie lasagna, the leftovers of which help ease us back into the workweek. Homemade dilly beans to go with Tuesday chicken and squash. Wednes-

day, a pot of mint tea to get past the hump. Thursday turns your garden into a salad bar, where everyone can compose their own bowl from all the ripe greens, herbs, veggies, and flowers the garden can yield. And fresh tomato, basil, and mozzarella sandwiches: perfectly easy for a Friday night dinner.

My kitchen is the place where I blend up cosmetics, herbal soaps and oils, kombucha, sauerkraut, healing salve, tinctures, butters, bitters, mustards, and vinegars—and the place where almost every pot of tea and meal originates. In the kitchen we are gardener turned pharmacist, alchemist, caregiver, prep cook, entertainer, and artist, melding backyard produce with spices from around the world, a grandmother's recipe, or our own creativity to craft meaningful and delicious connections back to the soil of our own backyard. Kitchen craft can be as simple or complex as we choose, but as gardeners, a comfort level in our kitchen can become our best superpower to foster family health, wealth, nutrition, and fun.

KITCHEN
GARDENS

I WAS TAUGHT TO ALWAYS USE the proper tool, and that each tool had its proper place. Kitchen gardens are designed as much to be a tool as a resource. They are a tool for healthy eating, for teaching household botany and resource management; and for untold centuries, their place was right outside the kitchen door for easy access. In them, we cultivate the nearest and dearest herbs and vegetables for our tables and medicine cabinets. We raise generations of children who help sow the seeds and run out to harvest sprigs for the soup kettle and teapot. Together, we learn the flavors, medicinal properties, and cultural needs of plants used in daily household botany.

Early books of husbandry and housewifery recommended raised beds in repeating four-square patterns radiating out along a path from the kitchen door. According to other primary sources and pictorial evidence, raised beds, approximately eight to twelve inches tall, were filled with a blend of rich compost and composted manure. Most were four feet wide and either four or eight feet in length. For ease of working and harvesting the rows of vegetables or greens within, one- to four-foot pathways of low-growing fragrant herbs, sand, or gravel filled the space between raised beds.

Beyond the kitchen garden, compost and wood piles, laundry and chicken yards, and privies would be further distant but still within the confines of the garden fence, within sight of the home. Historically in many places, a garden fence was required of the homeowner by law to protect crops from neighboring animals and people; the fence also created a safety net for chickens, children, and produce. We can still see evidence of gardens and homesteads like these around old cellar holes and often even in our own backyards—an errant patch of mint, thyme in a lawn, or daylilies that once bordered a path or foundation growing in a row.

In my own kitchen gardens, I usually build raised beds for perennial plants from stone; I have annual salad greens and veg in wooden raised beds, or just simple hilled earth with paths between. Root crops are in either wood or stone raised beds so long as they are one to two feet deep, to provide maximum clearance for deep-rooted plants like carrots, turnips, and salsify.

I can't say that I always put tools back in their proper place. My idea of a proper place is one that is convenient and accessible to me. For the same reason my kitchen gardens have not always fallen in line with prescribed method. Some have cropped up in a side or front yard, in pots on a balcony, in window boxes, or in a community garden down the street. Whatever inspires you to grow and eat your own veg is the perfect answer to what your kitchen garden should look like—a delicious labor of love!

LANGUAGE OF FLOWERS

BEFORE SCIENTIFIC METHOD, symbols and signs helped codify the natural world. As language grew more sophisticated, storytelling and literature animated the environment around us. Language speaks volumes, and human *nature* suggests that we are inextricably linked to the natural world. Our curious minds have tried to decipher how plants and animals communicate with each other and what perhaps they might be saying to us. In an effort to decode earthly and heavenly influences, we created systems of remembrance that helped us to pass our understanding from generation to generation. Especially in times before widespread literacy, the stars forming Orion's belt, dippers large and small, and hundreds of constellations moving across the sky marked the passing of seasons and years, and symbols helped us remember the attributes of plants and animals.

Native American animal medicines ascribed special meaning to significant animal sightings. Hawk as a messenger. Fox as protector of family. Deer bespeaking gentleness. Across many cultures, colors were associated with temperatures and emotions. Red as heat or anger; blue symbolizing cool; purple for royalty; yellow for happiness; white for purity. In the world of plants, we all recognized that the oak is mighty and the willow weeps. Longer than literacy and across cultures, a red rose speaks to passion. A white rose or the day's eyes of daisies blink wide-eyed innocence, and forget-me-not speaks its own name. Sometimes we rue the day, but sage speaks to wisdom, and borage brings back courage.

Flowers speak in poetry. The symbolic language of flowers has been recognized for centuries in many countries throughout Europe and Asia. Mythologies, folklore, sonnets, and plays of the ancient Greeks, Romans, Egyptians, and Chinese are peppered with flower and plant symbolism—and for good reason. Even the Buddha found enlightenment from lotus.

Yet by the Reformation, science and the written word was beginning to extinguish our need for symbolism. The ways we codified the natural world began to sound more like superstition and lore, and science began to replace other languages and symbols that nature had been known to communicate with. In becoming more literal, we lost some of nature's poetry. Many traditions that linked us to the natural world were almost lost to science and stoicism—that is, until the age of Queen Victoria, when a return to classic texts helped to reanimate the natural world and revive ancient seasonal observations and celebrations.

Victoria's reign also breathed new life back into the language of flowers, and our ears longed to hear that language spoken again as the old agrarian ways were dying out. As suddenly as industrialism began to break up our marriage with nature, romanticism took hold to keep the natural family together. Fairies, sprites, May Day, and dozens of the old holidays and holy days abandoned by the Reformation flooded back to rekindle our connection with the seasons and our kin from the natural world. In the art world, the Pre-Raphaelites, who aimed to revive the style of the Late Medieval Period, recaptured classic notions of beauty, romance, and tradition.

Meanwhile in America, new industries were being fueled by immigrants with Old World ways. Seasonal folk traditions transcended other linguistic barriers, and a formerly more puritanical nation joined in to celebrate our common roots in nature and her seasons. At the same time, publishing and printing in general became less costly and more mainstream; books, catalogs, magazines, and greeting cards began to find their way into homes everywhere. Along with this shift, botany and literacy among women was on the rise, and the first books that codified the language of flowers hit the gardening world by storm.

Besides romance, the Victorian era ushered in strict etiquette, including a near prohibition of conversations and flirtations between the sexes and social classes. So it could be said that the new passion for the old language of flowers was reactionary—a welcome means for covert communication in a repressive society. Language of flowers or floriography books and ladies' magazine articles ascribed meaning to every blossom and herb, and suddenly, young lovers were sending old-style tussie-mussies or nosegays. These small posies conveyed secret messages of love, scorn, loss, or hope that were deciphered by younger generations like today's ephemeral Snapchat posts, texting slang, and emojis. Perhaps we will step away from our blue screens and get back to the garden when we get our first woman president. Until then, please accept my posy of pinks (hope), nettle (defiance), and rhododendron (strength).

LAWN

I'M A BIT EMBARRASSED TO ADMIT IT, but I love my spring lawn. I relish its appearance. A clipped flowery mead, abuzz with life. I'm surprised by how much I enjoy mowing lush green stripes. Like a kid proud of coloring in the lines. There is a mindless reward in mowing. A walking meditation behind engine and blade. I am not an advocate of big monocultural lawns, but I am a grown boy acculturated to letting my mind wander while I create the appearance of order.

In my landscape, lawn has its seasons. From dandelion and strawberry to goldenrod and aster. Foraged teas, frisbees, blueberries, homebrews, salads, robins, and fireflies along the way. Verdant seasons, seasons dry as straw, and the majority of days between, adequate. Along with the seasons, the menu for bees and butterflies shifts from spring clover blossom to swaying summer daisies and fragrant milkweed.

Perhaps coloring in the lines isn't as important as providing a vibrant habitat for kids, wildlife, and pets and healthy drinking water down the line. I will continue to reduce the footprint of the lawn I inherited with more meaningful landscapes, but I will likely always keep a patch for spring—a happy habitat to celebrate lush times along the way. My weed wacker has been retired, and nature is calling on me to appreciate the wildness that grows up from the verge, the border plantings, and the meadows I encourage through benign neglect. I know that my lawn will soon succumb to the heat of summer; only the areas left as tall meadow will continue to bloom, seed, and feed life in the margins.

But for now, I will inhale deeply and relish the last of the spring green lines. With the arrival of July, I'll raise the mower deck and make way for the shaggy midsummer meadow, swaying with daisies and alight with fireflies.

MAPLE SYRUP

MANY CULTURES HAVE TRAVELED the world in pursuit of sweetness, but Native Americans found a source in the trees that grew all around them. Each spring, the sugar maple (*Acer saccharinum*) begins to flow with sap. In the colder parts of our continent, maple syrup can feel like recompense for long winters. A gift of caramel, liquid amber, that comes only from trees and people who know how to tap them. Smooth and comforting over breakfasts of pancakes or waffles. Sweet satisfaction stored and drizzled over bowls of late-winter snow.

Sugar bush, the old name for a grove of maple trees, gives a nod to the sweetness of maple syrup. Sap rising in earliest spring leads to syrup boiling in the sugarhouse, and a rite of passage that repeats its annual refrain.

The craft that begins with kindling wood, steam, and a passion for the sweet life goes back to North America's First Nations. Traditionally a tap or spile would be fashioned from a hollow branch, like those of elderberry and sumac. It would be driven into a maple tree to extract some of the sugars stored in the roots over winter. Ordinarily this sap would ascend into the tree's limbs to urge on early flower and leaf growth. Fortunately, healthy maples seem willing to share without showing any ill effect. Drinking this maple water as it collects has provided liquid spring refreshment for centuries, often when other freshwater sources were still frozen.

Certainly, our kin in nature taught people to drink maple water: every spring, I watch squirrels chew off the ends of branches to make them into sipping straws; I see dripping twigs swing low under the weight of the chickadees and nuthatches hanging upside down to drink. Maple water is not as sweet as syrup, but it has a hint of sweetness and maple flavor, along with minerally, earthy, and vibrant woody flavors that remind my body and palate that sap is rising and spring is near.

The whole process feels elemental. Collecting tree sap, starting a blazing fire beneath steaming cauldrons, and watching the vapors rise to carry away the excess moisture to render a thicker stuff. Reducing flow into essence. Water into nectar. When I stand by a kettle over an open fire or in a sugar shack, I imagine the process being carried out as it has for countless centuries as I stare into the billowing clouds of smoke and steam. It's a seasonal ritual for all my senses, much as it has been for people in generations before me.

Some people know only artificial breakfast syrups—fake butter flavor and corn syrup treacle in squeezable plastic jugs. But like all craft devotees, we make investments into products that are well made. Handcrafted goods that are healthier for us and the earth. While industry might find ways to chemically replicate flavors, or extract the greatest profit from a concept, we learn over and over again that they often come at a cost to our well-being and the environment. That's why I am saddened to see an age-old regional industry starting to use vacuum pumps to extract maximum flow from our trees. Suddenly trees that have been tapped for hundreds of years are sapped of every ounce of flow from root to leaf. Flow that no old tree can spare year after year. Tradition informs us about sustainable

balances that have been proven out over time. We can use science to help us maximize yield, but sometimes, it serves us well to let best practices and the wisdom of the ages serve as our baseline. Personally, I want to share sap with a tree, not rob it. It's like leaving enough honey in the hive for our bees to survive winter.

I also don't want my maple syrup running through miles of plastic tubing, as it increasingly does. Yet for the industrial bottom line, we drink the outflow like Romans from a lead pipe. A few years ago, I bought a jar of maple syrup that tasted just awful. When I returned it, I asked what could possibly have gone wrong, and they told me that the only thing that had changed was that they had begun to tap trees with plastic tubing. Instantly, the bad taste of sucking on a plastic hose to turn it into a siphon registered in my palate and memory, and it was exactly that taste that had imbued the maple syrup. Fortunately, these days in farmers markets, we can ask our farmers how they produce—and make our choices accordingly.

The Slow Food and other local economy movements remind me to slow down and taste the sweetness of life. Maple syruping time reminds me that everything is cyclical, that there is always sweetness if we remember how to preserve it. Our environment and communities always give back when we invest in them, and homestead crafts like maple syruping are helping a lot of my friends make ends meet.

Maple syrup may just represent the sweetest essence of the local food movement for me. Some people might gladly spend money to buy coffee out every day, or purchase costly industrial sweeteners, but for me, maple syrup is the sugar substitute of choice. It sweetens my morning coffee—a daily investment into a system I believe in. Local and artisanal foods are by their very nature fresher, sunripened, and made from a point of pride—not bottom line. And my morning coffee carries the vitality of sap rising, systems shifting, friends supported, health improved, and the sweet savor of a delicious revolution.

Agritourism and homestead farm stands have profited from seasonal maple syruping days, pancake days, and sugarhouses and farmers market stalls filled with maple candy, maple ice cream, and maple CBD sweets. Not everyone can farm, but every homestead with a sugar bush can contribute to local economies and supplemental income for our gig economy. There is a special alchemy that comes from gathering sap and boiling it down to syrup. Homestead craft enables almost everyone among us to tap into that deep-rooted tradition to make syrup out of rising sap that sweetens food, culture, and identity.

MAY DAY

ACROSS THE WESTERN HEMISPHERE, May is the month for flora, fauna, and gardeners to revel in the arrival of spring. In Celtic lands, Maia (the Roman goddess of growth and increase for whom the month is named) originally shared the May Day holiday with the celebration of Beltane ("bright fire"), which marked the midpoint of the sun's progress around the astrological wheel between spring equinox and summer solstice. On the eve of May Day, Beltane fires would be lit on hilltops across the land to ward off darkness and usher in spring.

Gardeners are known to rise with the sun, but May Day likely saw larger crowds, inspired by its associated traditions and superstitions. Upon rising on May Day, "rabbit, rabbit, rabbit" would be spoken out loud first thing, to ensure luck or fertility for the month. Leaving a branch of hawthorn at a friend's door was also said to bring good fortune. To ensure beauty, luck, and fertility, May Day dew from the leaves of lady's mantle was used to wash the face, rejuvenate the early riser, and restore a youthful complexion. "Strewing May," gathered at or before sunrise, would include every available flower and fragrant herb of early spring, though mayflowers (English hawthorn or trailing arbutus, our American mayflower), birch, and rowan were favorites. Medieval records note streets strewn so thickly on May Day that they could not be seen for the flowers.

A fair maid who, the first of May,
goes to the fields at break of day,
and washes in dew from the hawthorn tree,
will ever after handsome be.

—TRADITIONAL RHYME

Throughout history, May Day has ushered in the revelry, flirtations, and fruitfulness evoked by spring. Villagers would crown a Queen of the May and Jack-in-the-Green, who would lead the procession through the streets and cavort around the maypole. The maypole has lingered as a pagan fertility symbol; it divided north and south, and revelers danced around it, tethering the seasons together with ribbons originating from a wheel of life that historically crowned the maypole—over and under, over and under, making an artful weave. And the ever-turning wheel spins on, across the ages, in ancient churches, elementary schools, and backyard gatherings around the world.

May Day celebrations are an invitation to romp in nature; celebrate life and rebirth; plant seeds; eat spring delicacies; fashion posies, garlands, and bouquets; and drink in emerging fragrance and beauty after a long winter. May Day garden traditions include making May butter with sage buds; hanging May baskets on the doors of loved ones; eating a tansy or spring herb omelet; candying angelica and other edible flowers; and drinking May wine (with sweet woodruff).

Over time, the holiday has evolved. The day now also channels the hope and energy of spring into a campaign for human rights and a living wage for all workers. And it's still an opportunity to ring in spring, sow some seeds (and maybe some wild oats), and celebrate workers' rights across the ages! We might not get to strewing the whole street with flowers, but we can remember to send a May basket of violets or heartsease to someone we love.

MILKWEED

MOST OF US NOW KNOW MILKWEED (*Asclepias syriaca*) is a critical food source for migrating monarch butterflies—a botanical poster child for the interrelationship between pollinators and habitat. But through the years, milkweed and the native meadows it is synonymous with have been beaten back. With the advent of lawns and lawnmowers, it has been suppressed for nearly a century. Similarly, its designation as a "weed" speaks to the disregard and disdain that has been laid upon it. Yet more is revealed in milkweed's common name: its "milk" is a milky sap, similar to the latex found in a rubber tree. The bitter chemical composition of this latex is the reason that monarchs seek to eat the leaves and lay eggs on the plant. Monarchs can manage the bitter brew they ingest, but it makes them unpalatable to the birds that attempt to pick them off in flight.

Beyond monarchs, the milky sap of milkweed has a long history of human use, both as a topical antifungal and medicine to dissolve warts. Every other part of the plant was used, too. Once the fibrous stalks had died at the end of the season, many First Nations people would process them into strong twine and cordage. Generations of Native Americans, immigrants, and enslaved peoples boiled and ate the tender greens in early spring, before latex formed in the stalks. John Winthrop noted that women of the Massachusetts Bay Colony

used down from the pods to fill pillows and pincushions. In the Victorian era, women would ornament tea trays, wall hangings, and other art objects with milkweed down and affix the seeds to hatbands, to mimic snakeskin. During World War II, milkweed down was used as a substitute for kapok in life vests; and several companies are now insulating their down coats with it.

When I make wreaths for winter solstice, I use milkweed stalks among the greens. As the pods break open, they look like snow on the wind, but I also know that as they blow, they are sowing the seeds of future meadows and monarch havens. And the high floral notes and heady scent of milkweed is one of the sweetest introductions to summer that I know. The flowers are as intricate and beautiful as they are fragrant. So, instead of mowing this most important native plant down—have a whiff, appreciate its beauty and utility, and enjoy watching the pollinators and wildlife that you are encouraging!

NASTURTIUM

I'VE LOVED NASTURTIUMS since I was a child, and apparently, I'm not the only one. They have been the easy gateway plant for kids for generations, their brainy little seeds strange to look at but kid-sized and easy to grow. Like many before me, I felt like I was in on a special secret (code word "nasties"): I knew I could eat their spicy round leaves and orange and yellow flowers. Nobody else did that. You couldn't buy them in supermarkets, and other people just seemed to eat salads with bland iceberg lettuce and hard pink tomato slices. But I could grow and eat nasturtiums, and I was in good company: hummingbirds sipped nectar from them, and bees kissed them. Maybe I dove in for a little nectar too from time to time; it's still not uncommon to find my nose yellow from pollen.

I loved how the vines could grow up or down. Cascade over a stone wall, or climb up into the base of the old wisteria. The chipmunks would run up and down the wisteria's winding trunk to fetch, eat, or bury seeds. That was *my* clue to start gathering seeds for the following season, for more of those cress-flavored leaves, like miniature versions of Monet's water lilies, and sweet-peppery flowers. I became a lifelong devotee.

I also had an infatuation with Isabella Stewart Gardner and the museum she created, driving into Boston several times each year to see the changes in the courtyard garden. As spent clivia and acacia trees returned to the greenhouses, window boxes of trailing nasturtium vines would be brought to the courtyard—cascading green and orange curtains, year after year, since she began the tradition over a hundred years ago. I was then and am still enchanted. I also took design inspiration from the nasturtium vines that graced the pathways of Monet's garden. They turned footpath and arbor into a green haze dotted with a sunny orange and yellow glow that still motivates me to tuck nasturtium seeds into nearly every garden or planter where they can flow beyond ordinary bounds.

In the pages of old herbals, I learned you could pickle the green seeds or buds of nasturtiums to make poor man's capers. Now, as gardening seasons come and go, I have learned how to build upon my early season snacks and salads of leaf and flower—and mock-capers. High summer means stuffed nasturtium flower canapes, nasturtium butter, nasturtium flower vinegar, and nasturtium pesto. I have added nasturtiums to farmers cheese, spring rolls, and even summer cocktails. And they never fail to delight just as much as the flowers I ate as a child.

OLD WAYS

Walking.

I am listening to a deeper way.

Suddenly all my ancestors are behind me.

Be still, they say.

Watch and listen.

You are the result of the love of thousands.

—LINDA HOGAN

LINDA HOGAN'S OLD WAYS are those of her Chickasaw ancestors. Mine come from the landscapes of Italy, Ireland, and New England. They were emblazoned in our gardens, on our tables, and in our treatment of the land that supported us. They were passed down in recipes and prescriptions. In the way stones were fit together. In the way a dovetail joint expanded and contracted through hundreds of seasons. Old ways are how my hands learned to follow a pattern, save a seed, make a cordial, and trace the lineage that makes up every fiber of my being.

In centuries past, people's lives were harder in many ways. But most were also richer in meaningful experiences gained by living lives connected to the seasons and the elements. Lives more vivid for the visceral interaction of handcraft and better connected for the legacy of handed-down skills. Seeing these old ways as historical precedent and blending them with modern inquiry can help us to retrace familiar steps with new eyes. Look closely and analyze the memory embedded in fragments. We can find inspiration from the assured hands of a knitter, a calligrapher, a mason who came before us. Methods as simple as hanging out laundry or as complex as dyeing fabric with botanicals.

Although they can often be improved upon, old ways are guideposts that have helped us survive childbirth, illness, natural disasters, wars, and season after season of ordinary days. Why did particular herbs and spices become traditional medicines? Which heirloom seeds deliver the most flavorful, prolific, or disease-resistant produce? Old ways can shed some light, offering clues, like puzzle pieces, when we're fitting stones back into an old farm wall or letting our hand be guided by earlier cuts when we prune an old orchard tree. Old ways are the familiar artifacts that help us to explore an unwritten code, like talismans passed down to enhance and protect us. Ways so deeply embedded that they guide hand and heart to forge onward with minimal effort. Ways so old that only the wind and the trees can pronounce them.

No matter how dark the days, old ways can help us succeed and regenerate, like a woodland begins anew after a devastating fire. Old ways hold memories evolved through countless seasons of ice ages, hunting and gathering, agriculture and industrialization, feast and famine. As Rachel Carson reminds us, "There is something infinitely healing in the repeated refrains of nature—the assurance that dawn comes after night, and spring after winter."

For a multitude of reasons that vary all around the world, we have found ourselves at a necessary turning point in our relationship with the earth and each other. The gardeners' spirit knows why we have to take care of place. We know it from our very roots, and they are our tonic, our life blood—the water we drink, the soil we till, the heirlooms we grow, and the air we breathe.

Old ways can speak to us. They can lend us confidence and skills to do things like forage, but like any other language, their translation can make a world of difference. Old ways are nuanced, and they call upon us to be attentive to the details. Place, population, season—and how our decisions play out in the larger communities around us. At its core, respect is inherent in the old ways. If we poison the earth, if we do not apply lessons learned, we can poison ourselves; but if we pay heed, we can know no greater nourishment than the wild greens gifted by nature and birthright—nurturing medicine, like the lingering savor of a grandmother's chicken soup.

In an age when our waters and landfills are choked with plastic, old ways remind us how to make a wooden cutting board, and why it is best to rub it with garlic or lemon when we are done.

In a world with tidbits of food-like substances sealed in plastic to preserve shelf life, learning to pickle, ferment, and preserve foods that actually nourish our bodies becomes an act of resistance.

At a time when chemical fertilizers are sterilizing the land and toxifying our lives, old ways remind us to compost, companion plant, and cultivate perennials that draw up nutrients from deep below the earth's surface. They remind us how to eat a dandelion instead of pouring chemicals onto our lawn. In a world where we have sped up climate change and disease for the bottom line and profit margin, the answers we need (like dandelion) are blowing on the wind.

Old ways help us to roll up our sleeves and get things done. They help us to draw upon all the learning in the universe and the fifteen billion years of compost, in order to advance civilization and soil. It is our great fortune as gardeners and seed savers that we know how

to get our hands dirty, how to take spent soil and rebuild systems. In *The Same Ax, Twice: Restoration and Renewal in a Throwaway Age* (2000), Howard Mansfield concludes that whether it's rebuilding an old farm tractor engine or reviving the village model of community organization, it must contain an element of renewal—like fashioning a new handle for an ax broken so many times that there is little left of the original; yet its preservation communicates the spirit as well as the form of the original.

When I lived in Japan, I visited the Ise Jingu, a Shinto shrine some two thousand years old. But every twenty years, for at least the last thirteen hundred years, the most outdated building there is torn down and exactingly replicated. The process of regularly rebuilding the wooden structures, which would otherwise have been lost to the elements, helped preserve the original architecture; but even more importantly, it keeps the artisanal skills and processes (from harvesting and aging timber to timber framing and thatching structures) alive—often with the same lineage of artisans that built the shrine two millennia ago.

In contrast, on recent visits to Boston, I have watched an almost entirely new city being built on the old waterfront piers. Dizzying skyscrapers built at sea level on centuries-old landfill. I am keenly aware that craftsmanship did little to inform these plans. It would seem that nobody was tasked with creating buildings and landscapes that could be seen as living, breathing organisms, or that could endure decades of climate change. Clearly, in this age of excess, the old adage "a rising tide lifts all boats" no longer pertains to all; and these sinking ships may well be left bankrupt for the rest of us to contend with, like so many other outmoded dams and toxic waste sites from coast to coast.

We are not living in the most detail-oriented of times. Science can surely help us to advance, but first we need to know the right questions to ask. And all too often, the patterns most deeply ingrained in our history evade us.

Old ways don't require that we remain stuck in the past. Many are best abandoned and good riddance, but most simply remind us of patterns that have kept us resilient, helping us to survive a history full of change. Historian Howard Zinn urges us to remain hopeful, to remember the times and places where people have "behaved magnificently"; doing so "gives us the energy to act, and at least the possibility of sending this spinning top of a world in a different direction." The elders I've known who found ways to stay positive through changing times are role models to me. It's not that their lives were easy, but they never forgot all the ways that they were fortunate, and they made sure to acknowledge them as fervently as they worked for justice, equity, and beauty in the world.

There is nothing political about taking care of the earth. Every person knows it in their conscience; and old ways can help to guide us as we cultivate food, flowers, forests, and ecosystems, starting with the plot of land beneath our feet. They are living histories, and they are marvelous little victories. Open up the recipe box, the jar of seeds, and share the craft. Our work is to radiate out as far as we can, like ancient seeds carried on the wind.

ORGANIC GROWING PRACTICES

I HAVE NEVER SEEN ORGANIC gardening as a soapbox issue. Nobody wants to do harm to the environment, and everyone wants a chance to raise healthy kids in a safe place, with nutritious food. When we share the skills of gardening organically, we realize that most of its success lies outside of a consumer model. We can purchase costly organic yard products—or we can share skills and natural resources, "no purchase necessary": leave the leaves, build our soil, and let the birds and the bees get back down to business, while we grow our dinner and enjoy the show.

Each spring, I watch the families in birdhouses and nests outside my windows, and the adult birds going branch to branch, leaf to leaf, picking off bugs, caterpillars, and worms. It would seem that between the two parents, they are bringing a dozen beakfulls of inchworms to their young every minute. Sure, I could have sprayed chemical pesticides to keep the worms from eating the leaves on my trees, but someone, somewhere in a manufacturing plant is being exposed to those toxins, and effluent is likely seeping out into the waterways of less fortunate communities downstream. If I *had* valued perfect leaves above lives and habitats downstream, I likely wouldn't have had enough insects to feed more than a few nesting pairs in my yard.

Someday we shall look back
on this dark era of agriculture
and shake our heads.
How could we have ever believed
that it was a good idea
to grow our food with poisons?

—JANE GOODALL

When I look into my garden, I see millions of fascinating lives from forest floor to tree canopy. I revel in seeing shapes and colors, wings and fur, scales and shells, noses and beaks. I hear bird song, buzzing, rustling underbrush, and wind in the leaves. My breath is taken away by beauty and fragrance, not the holes in the leaves, and certainly not the smell of pesticides and fertilizers. And in all honesty, I have very few holes in the leaves on my trees, because of all the wildlife my landscape supports.

Years ago, I was recruited to be general manager for a chain of garden centers on Cape Cod. The owners, who had heard me subbing for Paul Parent on his *Garden Club* radio show, were concerned: within their lifetimes, nearly seventy percent of the agricultural chemicals formerly sold had quietly disappeared from shelves due to toxicity issues. However, the owners were smart business people, and our combined goal was not just to foster a healthier environment but to introduce organic alternatives that could be sold from the same shelves. No magic bullets, just traceable ingredients that were produced locally and internationally. Holistic, place-based systems, layering indigenous knowledge and sustainability science instead of consumer culture solutions. Like the first generation of any new product, these offerings tended to be more expensive, but the garden center had a dedicated following willing to walk past the industrial-sized bags of chemicals to purchase products that some of the more creative among them might have made in their own home and garden.

It was just over a hundred years ago that horticulture and agriculture began the shift from garden to lab, beginning with Fritz Haber, who received the Nobel Prize in chemistry in 1918 for inventing a way to synthesize ammonia from nitrogen in the air—a step that revolutionized the production of agricultural fertilizers. Thanks to research done by Michael Pollan and others, we can see more clearly the for-profit chemical attack on nature that was launched upon our lawns, farms, gardens, and kitchens after World War II. If we are what we eat, we're in trouble, ingesting, little by little, the byproducts of chemical warfare, glyphosates, and a lot of petroleum. Pull up to any gas pump, and there will be a sign reminding you to wash your hands: "Petroleum is a known cancer-causing carcinogen." So, I wash my hands, avoid plastics, and do my best to garden and buy organic.

After a century of chemical agriculture, it's no great surprise that the largest growing segment in 21st-century agriculture is organic. Consumer demand for organically produced goods continues to show double-digit growth, providing market incentives for U.S. farmers across a broad range of products. Organic products are now available in nearly twenty thousand natural food stores, and nearly three out of four conventional grocery stores.

It's wonderful to see organic agriculture finding its way back into the mainstream, but industrial agriculture continues to stretch the definition of organic to meet the bottom line, and sometimes the greenwashing is all too obvious. Fortunately, new generations of earnest market growers from local farms and cottage industries are rising to meet the demand for organic products made with integrity. Farmers markets and agritourism sites are not only helping us to know our farmers, but they are also providing opportunities to learn from one another as we revive old and apply new sustainability practices to right-sized, direct-to-consumer regional agriculture. And that is something to celebrate, when merely a decade ago there were virtually no alternatives to the industrial food complex left in our nation.

All my gardens have been organic. All meet, exceed, or inspire public standards of care and beauty, and all can be enjoyed as habitat gardens rich with wildlife. Mower decks are set high, and grass clippings are dropped directly back into the soil they will help enrich. Only the curious visitor's eye recognizes the mulch of leaves left as a weed barrier between perennials and shrubs. Hardly anyone notices that the topdressing in annual beds is really composted manure, or that established gardens lack in-ground irrigation systems. What they do see is thriving bee colonies and pollinators of every stripe, on pesticide-free plants. Lawns that look more like a flowery mead, rich with diversity, and clover that serves as the natural source of nitrogen to keep it green. What they will never see? Signs that warn about hazards to children, pets, or wildlife when the lawn company comes around.

We have a deep and proven history of organic gardening success, from First Nations to Martha Washington, your ancestors and mine. The past was not perfect, but organic gardening offers helpful historical precedents that can inform how we move into a more sustainable future as citizen scientists, linking up healthy habitats and green bridges around the world, one backyard at a time.

PERENNIAL CROPS

DO YOU HAVE ANY PASS-ALONG PERENNIALS? If you do, you know that they're often the most meaningful plants in the garden—welcome reminders of family and friends, living connections to people we still love. I cannot eat homegrown rhubarb, chives, or asparagus without recalling the joy my grandfather would express whenever he sat down to savor a perennial vegetable from his own garden.

A wonderful heirloom gardening practice is the cultivation of perennial food crops. In this age of local farmers markets, we have all grown to appreciate that the fewer miles our food travels, the better it is for our environment and our health. While many annual food crops can be direct sown, most are forced in less-sustainable greenhouses, with all their attendant chemical fertilizers and plastic pots, followed up by fuel oil and shipping. I am grateful for local greenhouses when it comes to getting a head start on tomatoes, eggplant, and other long-season annuals, but I also seize the chance to diversify my diet with many of the perennial vegetables that ornament my landscape and reconnect me with the pleasures of seasonal foods.

Perennial crops conserve labor and resources, and they extend the growing season by producing food that can be enjoyed nearly twelve months a year. The deep roots of well-established perennials

access moisture and nutrients when annual crops might struggle. These roots also push growth during the colder months, when pollinators and people are looking for food and few annual seeds would dare to sprout.

If you are shopping spring markets and garden centers, consider incorporating some of these favorite perennial vegetables into your landscape beds: asparagus, chives, walking onions, horseradish, Jerusalem artichoke, ramps, lovage, groundnut, bronze fennel, rhubarb, skirret, sorrel, salad burnet, chicory, and watercress. These and a whole range of perennial herbs and edible flowers add color and texture to your garden, diversity to your yard, and flavor to your life.

Think beyond your garden too! Small woody fruit bushes and berries are gaining entry back into our urban landscapes; and like perennials, these help us to eat seasonal foods that we might never taste fully ripe unless we grew them in our own yards. Some of my favorites include blackberry, raspberry, blueberry, currant, elderberry, gooseberry, beach plum, and cranberry.

What perennial legacy will you leave in the landscape when you plant up your wider yard this spring?

QUEEN ANNE'S LACE

QUEEN ANNE'S LACE SHARES its scientific name and history with *Daucus carota*, the vegetable garden carrot. As for its common name, legend has it that England's Queen Anne (1665–1714) pricked her finger and left a drop of blood on a piece of white lace she was working; and the flower of Queen Anne's lace does indeed look similar to a lace doily with a drop of dark blood at its center. Throughout history, such associations and memory triggers enabled generations to properly identify and pass down information concerning plants, which helped household botanists and foragers distinguish plants like Queen Anne's lace from dozens of similar white-flowering umbellifers—plants that ranged from edible to medicinal, abortive, and even deadly.

Queen Anne's lace (aka wild carrot) could be considered the ancient queen of all carrots. It appears as a common wildflower in fields, meadows, and pastures, yet it most likely arrived in North America as a garden crop. Among the earliest carrots brought to the New World were the white carrots known to and grown by English gardeners; they grew into full-sized carrots like those we eat today, but they were white.

A hallmark of heirlooms is that we can save their seeds, improving the plant with every generation of seed-saving we participate in. Specifically with biennial carrots, the traits valued were long, crisp, fleshy roots that would mature in one season and store well over winter. In addition, the colors selected tell stories of heritage and place. The earliest carrots known in the Middle East were purple or red; in eastern Europe, most were yellow. In the British Isles and Celtic lands, most gardeners knew only the white carrot—that is, until the 16th century, when the bright orange carrot favored by the House of Orange in the Netherlands made its market debut throughout Europe and the New World, including the Dutch settlement of New Amsterdam (later, New York).

Eurasian carrots with purple, black, or yellow roots from Turkey, Afghanistan, Egypt, Pakistan, and India are rich in anthocyanins (blue, violet, and red flavonoid plant pigments). Western or carotene carrots with orange, red, or white roots are the hybrid progenies of yellow eastern (Turkish) carrots and a subspecies of white carrot from the Mediterranean; they are rich in carotenoids (yellow, orange, and red pigments produced by plants and algae).

Any seed left without stewardship generally degenerates into its earlier form. If you pull up a flowering stalk of Queen Anne's lace, you will typically find a wizened and knobby white/brown woody root that smells distinctly of carrot. Since wild carrots are biennial, most blooming plants will have overwintered to the detriment of the root as an edible. Yet, if we were to select and save seeds of the fleshiest, longest, and straightest roots of Queen Anne's lace over several generations, we would once again be growing a white garden carrot. Fortunately, carrots are easy to grow from seed, and heirloom carrots of every color are now commercially available.

Queen Anne's lace stands as a reminder of those that have planted before us. A resilient adapter, with roots in the ground and seeds scattered to the four winds. In addition to actively self-sowing roadsides and meadows, the seeds of Queen Anne's lace carry a refreshingly clean scent, one that you might have noticed when handling carrots from the kitchen garden; not surprisingly, the essential oil of the seed is widely used in household cleaning products. As for the lovely flowers, I dry or press them to accent wreaths and mount pressed specimens on holiday cards and in windows, where they look like falling snowflakes—a fun floral craft that is easy to try in your home, too.

QUINCE

QUINCE WAS ONCE THE DARLING of dooryard gardens, nearly as common as lilac at house corners or entryway paths and, like lilac, serving as an early harbinger of spring.

European quince (*Cydonia oblonga*) is a woody shrub closing in on the size of a dwarf fruit tree; blowsy white flowers, their petals blowing in the wind like moths taking flight, give way to fragrant yellow fruit that ripens in late autumn. Flowering quince (*Chaenomeles speciosa*), a thorny woody shrub from China and Japan, is graced with exquisite flowers of red, salmon, pink, or white. Both species bloom in early spring, self-pollinate, and produce fruit with a balm-like scent that is both evocative and calming. But since it is not eaten as a hand fruit, quince is a rarity in supermarkets, and it requires preparation that few Americans are familiar with today.

Those familiar with quince most likely associate it with grandmothers who used it when making jams and jellies. When cooked with other fruit, quince provides a rich natural source of pectin that thickens preserves without excess sugar. Quince can feature in compotes, pies, and mince tarts and be added to poultry and lamb dishes.

This year, I have been enjoying a bumper crop of quince. I kept one on the dashboard of my car as an air freshener, and a large bowlful on my kitchen table perfumed the entire room for months. From harvest through the New Year, I continued to reach into that bowl to explore new uses, not all of which are complex. I learned to grate fresh quince into my oatmeal from an old Yankee gardener who proved that preparation didn't have to be fancy. I also add slices or chunks of quince to my teapot, along with fresh garden botanicals or a little green tea, for a heady, calming cup.

I candy wedges and chunks to use in pies and tarts (a pumpkin pie with a gingerbread crust, candied ginger, and quince being everyone's favorite). The syrup that results from the candying process makes incredible cocktails (Old Fashioneds and Rob Roys among them) and dessert toppings. I also put up several small mason jars of membrillo, the quince paste beloved throughout Spain when paired with Manchego cheese, and even experimented with turning quince

into a cordial. To do so, I filled large glass mason jars three-quarters of the way to the top with diced fresh quince, added a couple of chunks of candied ginger and some candied quince and its syrup to the fresh quince, and covered everything in brandy. After a month it was quite good, but by Christmas, with room to add honey, it became one of the best drinks ever to pass my lips.

In addition to being delicious when sipped, quince is a rich source of dietary fiber, vitamin C, copper, iron, potassium, and magnesium. Compounds isolated from quince have been shown to reduce blood cholesterol levels, prevent the development of certain types of cancer, and on a personal level—they mellow me out. So, cin cin: raise a glass to quince and good health!

RAISED BEDS

DOCUMENTARY, PICTORIAL, and archaeological sources provide evidence for raised beds back to at least Medieval times—wonderful woodcuts and earthly remnants of beds laid out in repeating four-square patterns, most about a foot tall and filled with soils best suited for each crop or ornamental cultivated: soil enriched with compost and manure for plants that require a heavy nitrogen load; thinner, sandier soils and deeper beds for root crops that demand good drainage. The beds were constructed of various materials—stone, brick, sheet lead, tile, woven wattle, and wood are all mentioned.

At Strawbery Banke, we recreated raised beds in their original locations based on the gardens unearthed onsite at the museum. The original beds were made with boards of wood, a plentiful material in the New World, held in place by wooden stakes to fashion boxes approximately four feet by eight feet. Evidence from one particular raised bed preserved a moment frozen in time: stakes had rotted off at the soil line and a wooden plank had fallen, causing rich soil to spill over into a garden path of river gravel.

With the excavation closed up, we planted reproduction beds three feet above the original garden. Despite the centuries that had passed, we could see that the raised beds had been positioned to capture the sun, hold heat and moisture, and allow for numerous plantings throughout the seasons. Pollen and seed analysis told us which plants had grown in and around that garden; and neighbors around the museum restoring their own houses began to donate seed packets they'd found in the walls of their homes, helping to recreate an even more complete picture of the heirloom plants that had been grown in that neighborhood.

In other exhibits at Strawbery Banke, I created more raised beds: several of these themed teaching beds were built using repurposed slate roofing tiles (we'd found evidence of this in another 19th-century garden in the community); one was a large oval raised bed of wattle for an ethnobotanical herb garden, to teach about the origin and use of herbs that came to New England from around the world. The same design features that make raised beds helpful for home gardeners make them perfect for children and other visitors to museums, public gardens, and seed-to-table gardens: they bring plant materials and labeling to eye level.

Raised beds not only look great, but they can also improve universal accessibility, making the work of seeding, weeding, harvesting, and viewing far easier for anyone tending them. This is especially important in senior centers and hospital healing gardens, where beds are best built at wheelchair height and depth or reach, and pathways are broad enough to provide ease of movement.

Raised beds are still best made from sustainable materials gathered close to home. They bring order to a landscape, hold fertile garden soil, and enable any gardener to weed and harvest more easily from the pathways that surround them. Above all, they evoke age-old patterns in the landscape, patterns that benefit household botany and horticulture across nations and continents.

RHUBARB

MOST PEOPLE WHO GROW RHUBARB (*Rheum rhaponticum*) can tell you where their plants came from. A grandparent, a generous neighbor, the person who lived there before . . . rhubarb is the quintessential pass-along plant. Yet, for generations, we knew rhubarb only as a dried root imported as a pharmaceutical laxative and purgative. The plant is thought to have originated in northern Asia, and the roots traveled the Silk Road to Turkey, Europe, and finally America. Philadelphia botanist John Bartram was growing medicinal and culinary rhubarb from seed by the 1730s; yet the first recorded recipe in English doesn't appear until 1807, when a recipe for rhubarb tart is printed in a cookbook by Maria Eliza Rundell.

Rhubarb hit U.S. seed catalogs in 1828, but several decades passed before the plant gained any real footing here—a slow introduction for a delicious vegetable, but like potatoes and tomatoes the leaves are poisonous, so it took time for the novice to trust that any other part of the plant could be eaten. It debuted in New England in 1870, when a Maine farmer introduced it (as "pie plant") at the Massachusetts Horticultural Society's annual spring flower show; and its subsequent rise in popularity coincided with the falling price of sugar, which made the tart rhubarb more palatable to the masses.

Where I grew up, rhubarb was a perennial favorite and one of the earliest harbingers of spring—a tonic that everyone seemed to value as a seasonal medicine eaten to drive away the long winter. Growing up, strawberry-rhubarb pie was one of the only seasonal foods we still recognized and appreciated, even if only for a few weeks each year. Pie contests, strawberry festivals, and nearly every restaurant, dinner party, or springtime community supper would have rhubarb accompaniments, from cobblers to rhubarb gin fizzes.

In the landscape, try using rhubarb as you might hostas or other foliage perennials. It produces some of the earliest spring growth in the garden, and as they unfurl, its large architectural leaves offer a bold and beautiful accent. The flower heads rise well above the parent plant, like curious plumes of cauliflower-like clouds; they blush to pink as they mature, finally forming striking brown seedheads reminiscent of buckwheat, to which they are indeed related.

It's always a pleasure to snap the first juicy stalk of spring, dip the end in sugar, and savor the change of season. From that first bite through to the first heat of summer that makes the foliage flag, rhubarb makes its way into many of the dishes that come through my kitchen. Beyond the ubiquitous combination of strawberries and rhubarb, fresh herbs like sweet cicely and angelica or spices like cardamom and rosewater help to elevate its flavor profile. I also use lots of raisins and prunes, or fresh and dried serviceberries (*Amelanchier*), currants, and elderberries to play off the tartness and help sweeten my rhubarb without using as much honey or sugar.

Best of all, I think of the old gardening friend who first passed it along to me decades ago and the perennial legacy it provides, spring after spring, from garden to table.

ROSEMARY

ROSEMARY will always have a favored place in my garden and in my kitchen. Not only do I love it as an edible herb and subject for landscape topiary, but its flowers are garden candy—sky-blue winged trumpets with a drop of sweet nectar, like a syrup of rosemary, in each. Rosemary uplifts poultry, meat, and bean dishes. It punctuates flavor when used as skewers for summer kebabs, and in winter, rosemary butter with orange zest is slathered on hot biscuits and under the skin of holiday turkeys bound for the oven. If my overwintered plants bloom, I make rosemary flower tea, by the cup or by the pot, to sip and enjoy as one of my favorite winter infusions.

> There's rosemary, that's for remembrance; pray you, love, remember.
>
> —WILLIAM SHAKESPEARE

In 2017, taxonomists and the genus *Salvia* claimed the rose of the sea as their own, *S. rosmarinus*. Yes, rosemary had been reclassified as a sage, and I tensed to think of my dear rosemary by any other name. Such a small matter, but my heart sank when the news reached me. Perhaps I wasn't as flexible or resilient as I'd thought.

I have worked in the field of horticulture long enough to see many plants reclassified. As a garden historian, I know dozens of plants by old and new names. *Hosta* was once *Funkia*, and some people continue to call it by that old genus name, a century after science articulated a new "surname." My favorite native, black cohosh, shifted from *Cimicifuga* to *Actaea* when I was in my thirties; it was not mind-bending for me. But somehow the shift from *Rosmarinus officinalis* to *Salvia rosmarinus* shook my underpinnings. Rosemary was the herb of remembrance, sage the herb of wisdom. I comfort myself with this: memory has always been an attribute of the wise.

Recent chemical analysis shows that both rosemary and sage serve the brain. Both perennial herbs are of similar stature, and upon reexamining the flowers, I saw that the flowers of rosemary are essentially miniature versions of the salvia blossom. Their leaves are what's different: rosemary with glossy green needles, sage with its silvery felted and tongue-like leaves. Most plants reflect environmental adaptability; clearly, rosemary retained small leaves to weather the winds and bright sun of the Mediterranean, whereas the larger leaves of sage evolved a light color, their silver-gray absorbing less heat from the sun.

Genetics are correcting the older, visual approach to classification, but our memories will remain. The root of science, after all, is devising a common language that helps us to better understand the world around us. My more adaptable self and history have also shown me that shifting terms do not negate the essence of all we've learned before. We too are part of the evolution of the botanical world.

A few days after I'd read the rosemary news, as I was making a pot of tea, an old intern called to say he was in the area and hoped to visit. I had to laugh when the next thing he mentioned was his dismay at hearing that rosemary had just been reclassified. As we talked, I was stripping the leaves off stalks of dried sage to make a pot of tea. Ironically, I was left with branches of small needle-like stems not dissimilar to the leaves of rosemary.

Clearly genetics will take horticulture in many new directions as time wears on. But I now recognize that the sage thing to do is to keep growing, learning, and seeing plant relationships more clearly with my own senses—and a gardener's power of close observation.

SAGE

DID YOU EVER WONDER why we always put sage in stuffing? Today we know that sage (*Salvia officinalis*) is an antibacterial herb. Our ancestors knew only that to ensure food safety, it was best to add an abundance of sage to the "belly pudding" (stuffing) and rub the bird inside and out with salt before filling the cavity (where we now know bacteria is most likely to form). For the same reason, sage was also a favorite ingredient in sausages. Scientific studies reveal that this and many classic flavor combinations are nothing less than proof that our ancestors knew a great deal about preserving food, food quality, and health.

Site sage plants in the hotter, drier parts of the landscape; an intertwined hedge or knot garden of red or purple sage with green, golden, and tricolored varieties is a particularly beautiful effect. As a perennial, sage is almost always available, so in addition to stuffing, I love to brown sage leaves in olive oil and butter as a quick sauce for pasta; the fried sage leaves make a delicious savory garnish when crumbled on top, too. Perhaps my favorite use of sage comes from the Medieval recommendation that eating sage butter in May will keep you healthy all year long. Every spring, I pinch most of the flowers and flower buds and chop or churn them into fresh spring butter, a condiment that when refrigerated lasts long into the season.

As a beverage, sage ale was popular long before hops ale, and it still makes a tasty brew! Sage is no stranger to my teapot either; I often blend it with fresh or dried berries (elder, hawthorn), cherries, or rose hips to make an astringent but flavorful tonic. *Salvia* translates to "life," so add some life to your kitchen and garden this season with some common garden sage.

SEASONALITY

SOLSTICES AND EQUINOXES mark the four movements in a celestial score. The older I get, the more I aspire to tap into the symphonic song of nature. To harmonize with the flow of seasons, the cycles in our landscapes, and the larger universe.

Winter solstice is the beginning of lengthening days and shortening nights; it occurs when one of the Earth's poles has its maximum tilt away from the Sun, which in our hemisphere occurs on or near 21 December. Ancient names for the winter solstice (from the Latin *sol*, "sun," and *stit*, "stationary"), including Longest Night, Midwinter, and Yuletide, frame traditions absorbed into the Medieval Twelve Days of Christmas and modern observations of Christmas and New Year—ceremonial time allotted to visiting and spending time with loved ones, eating, singing, wassailing, dancing, festivals, and fire—all urging on a return to the light.

Even though it represents the beginning of what is often the worst of winter weather, this solstice brings an end to winter's darkest days and welcomes in the growing light that helps us keep heart. Though bitter cold can sometimes disconnect us from nature and community, postharvest abundance and celebratory foods remind us to feast and share the perishable surplus of fruits, roots, winter greens, and holiday roasts. The same ancient holiday traditions

inspire us to carry boughs of evergreens into our homes and add the warming light of candles and fires to keep the connection alive through long cold nights.

At midwinter, the Twelve Days of Christmas offered weary gardeners time to visit and check in on one another. Sharing abundance when it was to be had and making kindness a ritual by commemorating the innocence of a newborn. If we saw need, we returned bearing food or firewood. If we saw loneliness or illness, we returned to raise spirits. Lighting a candle, raising a cup, singing carols, bedecking the halls, adapting a seasonal family recipe, and remembering our part in making life kinder.

I often go to nature to find peace and solace, but winter solstice reminds me to go out on a limb. Extend myself where I can, and take rest where I can. My agricultural roots remind me that even the earth must rest in winter, and that a new year offers a chance for introspection, connection, and renewal. Occasion to pore through seed catalogs and dream new landscapes into being. Time to slow down, observe, and reflect.

Spring or vernal equinox signals the beginning of spring in the northern hemisphere, marking the passage of the Sun across the celestial equator, as it travels from south to north. At the equinox (from the Latin *aequus*, "equal," and *nox*, "night"), around 20 or 21 March, Earth's northern and southern hemispheres are receiving the Sun's rays equally, and night and day are (nearly) equal in length. In fact, the spring equinox ushers in a long-awaited gardening season. For me, it means pruning orchard trees and roses, building wattle from the spoils, and listening to spring peepers sing out from vernal pools in the night. It's watching rhubarb and spring bulbs push up through the leaf litter (don't jump the gun and remove this vital protection too early!). Using the first cool, sunny days to work up a sweat, repairing walls and filling garden beds with the compost that winter turned to soil. Planting the heartiest of cold-weather crops— mache, arugula, spinach, borage, calendula, kale, collards, cabbage, parsnip, turnip, radish—to ensure a delicious spring follows. Making teas and salads from the first perennials and self-sowing greens. Using the growing hours of sun and heat to follow the season and plant more tender annuals when they will succeed. This equinox also means vestiges of the ancient rites of spring, rebirth, and renewal celebrating the pagan goddess Eostre, Passover, and Easter— holidays traditions that make good use of the abundance of early spring herbs, eggs, and dairy.

Summer solstice, around 21 June, marks the time of the longest day and the beginning of summer. In the Northeast, it ushers in prime growing weather and a magnificent rush of flowers and produce to enjoy and share. In centuries past, Midsummer's Eve was seen as one of the times that the fairies were supposed to come out and dance; today a fairy house movement plays off this theme, inviting children lost in the pretend world of blue screens to take a closer look at nature and play for free in an outdoor world of imagination. For families it's a time of cucumbers, summer squash, and zucchini. Garden salads, strawberry shortcake, picnics, and summer vacation.

The autumnal equinox falls between 21 and 24 September, as the Sun heads southward, crossing the celestial equator. In Celtic tradition, this equinox was celebrated as Mabon, which featured the building of an altar laden with harvest fruits and vegetables—an offering to the goddess or mother earth and a cue for people to count their blessings. In the Old and New World, harvest season and cooler temperatures typically led to more foraging, butchering, and baking. Late-harvest grains were made into seasonal breads, cakes, and beer; nuts and late-season fruits were turned into pies, cider, and brandy. Harvest festivals brought people together around a common table to celebrate the abundance of the season, the landscapes they shared, and community.

Gardeners and farmers are perhaps tied to these cycles as much as any of us can be, living in the modern world. And as a gardener still crafting a life and livelihood from the land, I am touched and heartened by our human impulse to mark the seasons together with earthly and celestial rites and rituals that help us celebrate our circles around the Sun.

SORREL

SORREL (AKA FRENCH SORREL; *Rumex acetosa*) is one of my favorite heirloom greens, its leaves larger and milder than those of sheep sorrel (*R. acetosella*), its wild granny; but both epithets—*acetosa* and the diminutive *acetosella*—refer to the citrusy, sour acids in the arrowhead-shaped leaves. As a perennial, sorrel (from the Old French *surele*, "sour") is among the earliest and most tartly flavorful spring greens. The leaves are chartreuse, though often flushed with red from cold nights when they first emerge. They rise above ground in tufts that push aside leaf litter to follow the sun.

After a long winter, these first nibbles from the spring garden offer lemony refreshment for the taste buds, liver, and kidneys—giving sorrel the reputation as a tonifying herb for spring. If left unattended (or you are not using it enough) as the season wears on, the leaves weather, growing astringent and bitter, so at least once a month the whole plant should be cut to the ground so that a new flush of tender leaves will grow back, a cycle that you can continue from earliest spring through hard frost.

Sorrel is a delicious source of calcium. The leaves can be added to salads, or processed into a lemony green sauce or pesto, as beautiful as it is tasty. The milder new leaves also make my favorite, sorrel soup. There is no better way to serve up a taste of place and season

than a sorrel soup dressed with fresh Solomon's seal flowers. I once had it paired with steamed asparagus with chive blossom butter and a thicker sauce of sorrel and bronze fennel leaves over whole roasted fish from local harbors—a simple meal to mark the establishment of an EcoGastronomy program at the University of New Hampshire. The community has now graduated several classes of EcoGastronomy students, created farmers markets in nearly every town, supported agritourism laws and farms, and fostered farmer-chef connections. The vibrant local economies that are the natural result of all this recall the fresh taste of sorrel to all who gathered around the community table, that memorable day.

TEA

I GREW UP WITH A MOTHER who lived in the moment—at least when she had a cup of tea. Every time she raised her cup to sniff the vapors and take her first sip, she would exclaim, after a beat, "This is the *best* cup of tea." Her little mantra of appreciation.

As my day goes on, I love my time with a cup of tea too. Usually Earl Grey in the morning for the gentle caffeine kick and the sweet citric note of bergamot. I may go on to green tea in the afternoon, but most often I just make a pot of herbal tea to sip throughout the day. As an herbalist, I look at each cup as a dose. Each teapot as a prescription of season and place, a complementary medicine from my garden pharmacy. "First, do no harm" is attributed to the Greek physician Hippocrates and a pillar of our modern health care system. I like to think of my garden and teapot as the first line of defense.

At the height of summer, I pick bee balm flowers and fill a teapot. Sometimes the hummingbirds are so intrigued by the scene that they continue to sip nectar from the flower stalks in my hands. I either make sun tea, or I let boiled water drop a few degrees and pour it over the leaves and flowers. I inhale the first vapors and put a lid on the pot to keep the distillates within the pot.

This type of tea was first known as an infusion or tisane. A typical infusion would be made from the tender aerial parts of the plant—leaf and flower infused into hot water. If the woody parts of a plant like the root, bark, twig, seed, or hard fruit are to be prepared, they would be simmered into a decoction in order to extract the flavor and medicinal attributes.

Often when I design a garden space, I think in terms of planting a garden of allies. They might be balms for the spirit, or remedies for ailments that I am prone to. For instance, there is scarcely a person in my family with a surviving gallbladder, and from earliest childhood, I remember my grandmother and mother preparing a dose of bay leaf tea, perhaps my first foray into the positive effects of herbal medicine. Needless to say, I always have a bay tree growing in my garden.

For teas, I have often grown and enjoyed chamomile for upset stomach. Mint for sluggish digestion and to open my sinuses. Haws (hawthorn berries) for heart. Fennel to support diet. Rosemary leaf and especially flower for my spirits. Sassafras for a spring tonic. Valerian to induce sleep. Thyme and sage for their antiseptic and antibacterial properties and flavor. And throughout the season I take inspiration from what can be gathered fresh. I once had a garden that was relatively orderly except for the constant upstart lemon balm plants. Over time, I pulled up entire plants of this herb, snipped off the roots, and filled the teapot with balm. As I inhaled the scent and sipped the soothing green liquid, I came to accept that it was nature's way of reminding me to pause, inhale deeply, and chill out.

Lemon balm was an herb for an earlier time: never again has it been so prolific for me, and it doesn't dry well—so on to the next. There are so many plants to enjoy and experiment with. I head back into the garden with some sun tea, and I dry and preserve all the herbs I can for the winter months ahead, when hot herbal teas serve as my primary means of hydration. A dose of tea, a spot of medicine, frontline wellness from the garden, and a cure for anything that ails you. And as it did for my mom, the soothing ritual of sipping a cup of tea brings me back to center to see the world anew. In life, there are very few things that a steaming cup of tea and a hot bath can't cure.

THANKSGIVING GRACE

LATELY, I'M HAVING DIFFICULTY approaching Thanksgiving with grace. Struggling with the state of our union before we even sit for dinner. So, I attempt to move beyond political battlefields and remember that food can be a healing medicine. I think of my Irish grandmother, who wanted only a wreath of bittersweet at her graveside—"because like life, it is both bitter and sweet." She was also quick to remind us: never forget the sweet. Certainly, this holiday has roots that offer both sweet and bitter cues for reflection. As bittersweet as life can be, I still hold Thanksgiving as a persistent challenge to reach across the table, a reminder to celebrate the sweet. I want to relish life, to learn from history, to remember the millions of deeds that come from love, not hate, and the ordinary acts that bring us together over tables.

Life can be easier when we offer up an appreciative hug, a kind word, a genuine olive branch. Not humble pie, but local parsnips, cider, cheese, and beer. A jar of garden preserves, an old family recipe for cornbread. I'm slowly learning that the stewardship of shared habitat is perhaps best learned communally around a potluck table. Snapping beans in good company. Peeling squash in a stream of

The ordinary acts we practice
every day at home are of more
importance to the soul than their
simplicity might suggest.

—THOMAS MOORE

sunlight. There is healing that comes from cooking together. Deep nourishment that comes from forging solutions around to the table. Day-to-day living that provides common ground for the seeds of our shared future.

So add an extra leaf to the table. Keep the tradition of breaking bread with the stranger, reserve a place for that relative you struggle, lately, to welcome wholeheartedly. We are all products of our times, and our aunt's Brussels sprouts. Remember that collards rock. Find the thread. Revisit meaningful recipes, scratch bake in good company, teach a child to knit, make apple butter together. Slather it on. Cut open another pumpkin and sort seeds to roast, and seeds to plant.

Try to grow beyond the familial thicket and graft onto better root stock, when you can. Gathering around groaning tables, we can let veggies and love be the things we serve up in excess. As we peel, chop, and roast, we add ourselves to the mix of countless generations that have pondered, over pumpkin pie, how to grow a better squash and advance civilization. Clasp hands or knead dough, and get back to the earnest work of living. There is still time to plant next year's garlic, fill our cupboards, share some cordials with kindred spirits, and replenish ourselves for the cold season ahead.

Last year, the calendar on my cell phone made no note of Thanksgiving—but there was Black Friday, highlighted as a holiday. Wampanoag and Pilgrim sensibilities tell me to continue to mark a day set aside for giving thanks, flawed as it is, and to turn off the blue screens of consumer culture selling pharmaceuticals and plastic holiday novelties made in China. I'm not inclined to let my phone define my holidays. So this year I'll spend the day exploring nature with visiting kin, or just stay home, feet up, and enjoy some Black Friday pie.

TONICS

We need the tonic of wildness.

—HENRY DAVID THOREAU

TRADITIONALLY, tonics were seen as medicinal substances taken to give a feeling of well-being or vigor. But anything invigorating physically, mentally, or morally could now be said to be a tonic. A kind greeting can be a tonic. So can nature, gardening, art, and music. All refresh the human spirit.

Health. The word comes from the Old English for "whole," as in the state of your whole being—mind, body, and spirit. The old herbalists used herbs to address the root causes of illness; their garden botanicals were a prescriptive and dietary means to balance the whole system and maintain good health. In contrast, much of the focus of modern Western medicine has been on suppressing symptoms, with pills. But household botanists understand that the foods we eat and the tonics we sip support our overall health and well-being, before we need a capsule.

I do my best to stay well by finding ways to support my health from my garden. The tonic of fresh air and exercise, the tonic of working in my garden, the tonic I sip to refresh my spirits, the spring tonic that renews after a long winter, or the gin and tonic I sip on a sultry summer night.

Cultures with limited sources of fresh food in winter often have traditions regarding the restorative and curative powers of spring's earliest plants. Spring tonics were seen as a means to restore vigor after a long winter's diet of mostly preserved and less vital foods; they could be as simple as eating a particular dish of greens, sipping on an herbal tea, or partaking in a wide range of blended botanicals that were used to increase stamina, keep the body regular, or invigorate mind and body, as some use coffee or chocolate today. As a gardener, my spirits soar when I harvest the first greens of spring. I know that I am imbibing all the vibrant life force and energy stored beneath the ground through winter, whether by adding chives to cheese or making a rhubarb cobbler or lovage soup. It's wonderful to stir your senses with an early spring salad of violets, arugula, and sorrel cultivated in your own backyard.

Early settlers also delighted in eating the first "knots and buds" (tender leaves, buds, and flowers) of spring. Much as we do today, they found great satisfaction and rejuvenation consuming these vital outgrowths of warming earth and sun. They drank sassafras tea, ate beets and spinach to improve the blood, chewed on parsley for the vibrant green flavor that chlorophyll imparts. These settlers were firm believers in the tonifying effects of spring greens, which were said to stimulate digestion, cure scurvy, combat rheumatism, and ease constipation after a long cold winter of preserved foods and relative inactivity.

One beloved spring edible is the ramp (*Allium tricoccum*). First Nations and early settlers valued these ephemeral native leeks as both a spring tonic and a culinary onion. Like garlic, these members of the onion family are used to improve health, cleanse blood, and protect people from illness. They have scientifically proven antiseptic and expectorant properties that can assist with colds and virus. Research

also shows that ramps supply vitamin C, minerals, and allicin, which has antimicrobial and antithrombotic (blood-thinning) properties—all attributes that make it an enviable tonic for early spring.

Because the earth needs the tonic of co-conspirators too, I have learned to cultivate both ramps and fiddleheads in the woods and gardens around my home, doing my part to strengthen, instead of deplete, wild populations. But if these native plants are scarce, my Italian grandmother's simple dish of dandelion greens is a spring tonic fit for any table. They are known as an edible liver and kidney tonic, and there are always more than enough to go around.

Not all tonics, however, were homegrown or foraged; many were created as pharmaceuticals. These apothecary tonics and bitters, ubiquitous from Eurasia to the New World, reached their pinnacle in the 19th century, as people fell out of sync with the tonic of nature and found themselves working in cities and factories. They were intended to provide vigor for weary workers and aging bodies. Ginseng (*Panax quinquefolius*), sarsaparilla (*Aralia nudicaulis*), and their Asian relatives were used as male tonics to increase libido. Blue cohosh (*Caulophyllum thalictroides*) and motherwort (*Leonurus cardiaca*) were used as female tonics for women's health; and nerve tonics made with nervine herbs (skullcap, chamomile, valerian root, hops, lemon balm, St. John's wort) were tinctured and popularly prescribed to women with melancholic and choleric tendencies—typically depression or boldness, no doubt brought on by being an intelligent, spirited, or sexual woman in the buttoned-up Victorian era.

Their male counterparts, who were generally freer to carouse, often self-prescribed with alcohol. In the bars they frequented, a new generation of cocktails was made using bitters from mugwort, passionflower, wormwood, angostura bark, orange peel, and other botanicals—tonics intended to support liver and kidney functions that were undermined by the consumption of alcohol in the first place.

As the temperance movement suppressed public consumption, producers found legal loopholes in the historic method for tincturing herbs in alcohol to preserve them. Mass production, along with new means of advertising and delivery, helped a generation of Victorians to disguise booze as a botanical wonder drug. In the process, they profited from bottling swigable doses of alcohol-based tonics that could be purchased without stigma. Unfortunately, many of these patent medicines and cure-all tonics became vehicles for delivering addictive doses of alcohol and narcotics to unsuspecting consumers and families.

Meanwhile, overseas, tonics were taking another turn. Quinine powder prescribed to combat scurvy was so bitter that British officials stationed in India and other tropical outposts mixed the powder with soda and sugar. By 1858 the first commercial tonic water came to be a staple medicinal—with an alcohol chaser. That pairing was the gin and tonic. And for those drawn to the temperance movement, a new generation of syrupy tonics like sarsaparilla, root beer, and cola were designed to appeal to wholesome industrial workers while temporarily jacking them up on cocaine, opium, or caffeine. Feckless producers were eventually reined in: the coke in Coke was removed, the lithium in 7-Up was struck down, and only the fizzy, sugary pablum remained.

For better or worse, all these shady tonics helped usher in the age of modern medicine and medical reform. Unfortunately for gardeners and domestic botanists, the baby was thrown out with the bathwater, and for almost a century, herbs and herbalism were demeaned as the stuff of old wives' tales—until herbs were finally analyzed scientifically for medicinal use.

I could say I use caffeine like a tonic at times. Certainly, the early Coke ads promoted it that way. But mainly I see tonics as any refreshment from my garden or spice drawers that supports my well-being, whether that means breathing in the scent of a rose, sipping a cup of restorative ginger tea, or drinking beet kvass.

Any week, in every season, is improved by tonics. Friday night I was uplifted by some spruce beer and an herbal gruit at my local brewery. Next morning's tonic evolved from a gift of frozen homemade wild grape juice: I brought the thawing juice to a high heat and turned it into an oxymel, simmering it down with cider vinegar from a neighbor's orchard and local honey. Suddenly, in the middle of winter, I was sipping small doses of sweet, piquant, late summer from a spoon, nectar from the vine, and a tonic that tonified my gut; with the addition of vinegar, it acted like a roto-rooter for my arteries and provided me with a much-needed megadose of vitamin C. Saturday night's tonic came in the form of a Sazerac (said to be the oldest cocktail in America) made with wormwood absinthe and Peychaud's Bitters. Sunday was a good day for a dose of lemon balm in the teapot, and Monday called for ginseng to wake mind and body. Tuesday brought sunshine, a tonic all its own. Wednesday, it took a strong dose of coffee to bring me back into the world, and some kombucha, like a probiotic version of the grape oxymel, accompanied lunch. Thursday, I drank a Moxie with my sandwich to get through the afternoon. In New England, we still call soda tonic. Perhaps, as the old herbalists would say, it's just in our blood.

VALERIAN

VALERIAN (AKA HELIOTROPE; *Valeriana officinalis*) grows tall and skinny, sweet and drowsy, sending up tall, cottony flowers like clouds atop chopsticks. These highly fragrant, light pink flowers tend to rise above peers and reseed readily around the garden. Vast fields of valerian grow along the old shore routes of coastal Maine, immigrant heirloom ancestors in a region where it's been cultivated since the 17th century. The plants we see today are just as likely to be descendants from the garden of a sailor's wife or Shaker seed house as escapees from a later Colonial Revival garden. Generations have had easy access to the medicinal roots by simply thinning excess plants to harvest roots for a sedative tea.

John Gerard and many other early herbalists noted valerian's usefulness for treating insomnia, and science has validated that use. It's the equivalent of a chill pill in a teacup or capsule, relieving nervous tension and aiding sleep—and it doesn't have addictive properties unless you are a cat (cats are as attracted to valerian as they are to catnip). Most people who take valium likely don't know that it

was derived from valerian. Sadly, and perhaps not surprisingly, the chemical used to replicate valium as a patentable pharmaceutical has addictive properties; valerian root does not. Ironically, the very herbs that inspired nearly seventy percent of our patent medicines are routinely called into question as primitive or unsafe. But we are finally beginning to learn how the marketing budgets and political lobbies of the pharmaceutical industry can buy influence.

I never prescribe what others should do or take; we always need to research, exercise caution, and build firsthand knowledge since ten percent of the population seems to be allergic to something that you or I routinely eat or dose with no ill effects. But personally, I trust the valerian root from my garden or forage. Occasionally, to induce sleep or sedate my over-thinking brain, I add it to a teapot with complementary herbs—lavender, hops, any calmative or soporific plant. In years when more valerian runs through my garden than serves the space well, I pull plants up by the root for wellness. I usually make fresh tea when plants are in flower or leaf, and tincture the roots after frost, when the strength of the plant has moved underground.

Valerian has many wonderful properties, but the old-gym-sneaker smell of the dried root is not one of them. That's why mixing complementary herbs and making herbal blends has been part of the herbalist's craft for centuries. So, like centuries of herbalists before me, if I am brewing valerian root tea, I blend the decoction with season, flavor, and purpose in mind. In winter, I might offset the "fragrance" of valerian root with fresh wintergreen to add a deep-forest, minty essence (and a mild analgesic) to the brew. In summer, fresh lemon balm uplifts the tea's flavor and fragrance and ups its calming effect. If I want my night's sleep to be filled with dreaming, I add mugwort—and sipping a cup in the moonlight, whatever sliver there may be, I ready my weary head for bed.

VICTORY
GARDENS

AT THE HEIGHT OF THEIR SUCCESS during World War II, Victory Gardens provided nearly half of U.S. produce. In a time of adversity, our government refocused and united the nation, partnering with defense departments, cooperative extension services, and civic groups; investing tax dollars, resources, and personnel to retrain citizens to save seeds, raise crops, and feed a nation whose farmers were fighting overseas. First Lady Eleanor Roosevelt turned the White House lawn into a model Victory Garden and persuaded both skeptical officials and the nation's families to take up spades, shovels, and hoes, dig up unproductive lawns, and plant home Victory Gardens to win the peace.

The community garden movement was launched for those who did not themselves have their own land to plant, and those neighborhood plots became places to advance garden training and teach food preservation, while simultaneously instilling the values of self-sufficiency, community engagement, and collaboration. Even many of the public parks, intended to provide greenspace for post-industrial urban areas, were turned into shared gardens; many live on as community gardens today.

Suburbanization and industrial foods were still new to America, but in a few short generations, what had been a majority agricultural nation was already forgetting how to farm and garden. Trains, streetcars, electricity, automobiles, world war, economic depression, and urbanization had recreated the American way of life. Post-Dustbowl citizens were enamored with the seasonal defiance and stability of canned goods, and many had learned to forgo flavor in favor of shelf-stable foods. Agriculture continued to some extent in the Midwest, certainly, but with farmers drawn into the latest war effort and without local farms across the nation, America could not feed itself without a major change.

With the Victory Garden movement, the nation turned on a dime. Government agencies and universities produced movies, garden plans, booklets, and training programs that helped families plan to feed themselves from their own yards. Ironically, while canned goods were sent overseas to the troops, many in younger generations relearned the unparalleled flavor of sun-ripened tomatoes and came to recognize the distinct difference between fresh produce and canned peas, broccoli, and spinach. And older people who remembered how to live from the land joined with the women and youth that remained behind to create a brilliant new homefront workforce for every garden.

Many of the young men who were recruited for service from impoverished communities (or what we might now term food deserts) were found to be malnourished. To address this, new educational programs focusing on the science behind health and nutrition were taught along with lessons in planting, harvesting, preservation, and home economics. Posters even reminded families that Victory Gardens would allow them to grow "vitamins" outside the kitchen door.

In every community across the country, Victory Gardening was taught as a craft: canning, cellaring, and composting; seed sowing and saving; fruit production and small-yard farming efforts like raising poultry and livestock. Similarly, school garden programs and school lunch programs were institutionalized in order to further these lessons across generations.

While at Strawbery Banke, I had the opportunity to rehabilitate an original Victory Garden in the neighborhood that eventually became the museum, beginning with remnants of the original fencing, stone paths, and laundry umbrella, and botanicals like yucca and New Dawn roses. Oral histories and pictorial evidence filled out the plant selections and planting patterns. We learned which plants were valued in our climate and community at a time when many seeds were still open-pollinated and regionally diverse; in some instances, we found heirloom legacy that went back to Native American seed selections. The annual yield from the garden was tremendous, but so were the learning opportunities. Many of its lessons were eagerly applied as new community gardens, school gardens, and seed banks were established. Nearby restaurants, garden centers, and farmers markets began to cultivate and use many of the same plants, once they understood the regional importance of the seeds, plants, and recipes.

In recent years, and times of war, we have been told that our patriotic duty was to get out and spend. During World War II, selfless volunteerism and contributions of valued—not hoarded—household resources went beyond patriotism; they were recognized as the best way to survive as a nation. We willingly gave over public and private lands for food production, shared tools, and learned to "use it up, wear it out, make it do, or do without." We also learned that tending a garden big enough to feed your whole family took the whole family, and that canning garden surplus was far better done in good company than on our own.

The Victory Garden movement is a reminder that when we are not suckered into priming the pumps of consumerism and industry, we can reinvest in the health, education, and welfare of our families. In the face of new challenges, we are planting the seeds required to bring a population together around common aspirations. And in the process, we are also remembering that it's a whole lot more fun, instructive, and delicious to snap and can bushels of dilly beans with friends and family.

WATTLE

WATTLE IS MADE BY WEAVING or interlacing twigs, branches, or saplings to form a fence, enclosure (a raised bed, for instance), or garden structure (arbor, trellis). The process, which dates back to the Neolithic age, is essentially like weaving a basket from the ground up: pliable, slender stalks are woven between strong upright branches staked into the ground. Wattle can be made from long, straight branches pruned around your yard, or by harvesting the water sprouts from orchard trees.

Like basket making, wattle is an essential craft from the landscape. In England and Europe, hedges, coppiced, and pollarded trees were managed as renewable resources; the branches collected were commonly used for household projects and fuel, since Medieval European landscapes were largely agricultural and supported few large trees. The first immigrants came to North America with these crafts and skills, and in our earliest homes, wattle and daub served like lath and plaster, or early alternatives for sheetrock and insulation; however, in colonial American landscapes, timber and stone were available in abundance and usually provided the material for the fences and walls that divided fields, pastures, and neighboring properties.

With so many of us living in urban or otherwise built-up land-scapes, wattle once again provides an opportunity to use place-based materials to craft sustainable garden structures. Best of all, it is easy to grow everything you need to have a continuous supply for gardening projects and repairs. You can create an annual harvest for these purposes by treating your trees and shrubs to coppicing (cutting back to ground level) or pollarding (removing their upper branches); these pruning techniques stimulate growth and promote a dense head of branches for firewood, craft, or construction mate-rials. If coppicing or pollarding trees seems too intimidating, it's easy to plant shrubs with a long history of use as a withy in thatching, basketry, and gardening.

Some of the easiest plants to grow and weave include alder and red or yellow twig dogwood, as well as several willows (*Salix acutifolia*, *S. daphnoides*, *S. ×mollissima*, *S. purpurea*, *S. triandra*, *S. viminalis*). All are easily managed in hedges, rain gardens, or areas where you might want to create a natural buffer with useful plants. Since these are moisture-loving plants, choose your planting area well. Clear and prepare the site and take cuttings or slips from mature specimens before they leaf out. Make sure that the cuttings are long enough to slide at least a foot into damp spring ground, where they will strike roots. Within three years you should have a lively hedgerow fit to harvest. Withy grows quickly and responds so well to harvesting that you may want to add another garden craft to your skillset and take up basket making to utilize the renewable resource you planted!

Quite often wattle fences take on a new life altogether. Fre-quently, in the first year of making a wattle fence, the uprights (or, alternatively, interlocking hoops tucked in the earth to line a garden or path edge) will sprout leaves and sometimes even flowers—an endearing and often strengthening feature of any wattle edging or fence. In fact, I had cast iron "woven wattle" edging replicated for the Victorian children's garden at Strawbery Banke: having watched several generations of wattle edging sprout, it was a delight to find that Victorians (who commonly used orchard trim to make wattle) had captured that tendency, complete with apple blossoms, in cast iron, too.

Wattle will not last as long as its Victorian metal counterpart, but a well-built wattle fence can be capped off or mended annually in a way that will make it last for many years, as long or longer than the plastic edging that so frequently winds up somewhere between our gardens and the landfill. During the Victorian age, materials were expensive and labor was cheap; now we have learned that cheap materials also come at a cost.

Wattle offers a simple reminder that craft can help turn utility into art. I find the process deeply satisfying. The warp. The weft. Weaving together fragments of the landscape I love. In and out, over and under. A meditation on landscape and sustainability—and a chance to play with sticks again as an adult.

WREATHS

WREATHS (from the Middle English *wrethe*, a twisted garland or ring of leaves and flowers) have been used ceremonially for centuries. Since ancient times, they have symbolized eternity, because the circle or ring shape has no beginning or end. A linear path merely takes us from beginning to end, but the wheel of life symbolized in a wreath reminds us that we are always most vital, engaged, and alive when we accept our current phase, create forward movement, and celebrate the season at hand. Entwined botanicals form an artful intersect between natural and domestic worlds—and turn a front doorway into a portal bridging the two.

Gathering backyard botanicals for holiday decorations can connect us to age-old traditions, celebrations born of the season. As we walk outdoors, taking a long look back into the garden and forest floor, the bones of the landscape are evident. We can see where to prune and where we could improve on plantings in the year ahead. Taking a walk, gathering natural materials, making a wreath—all become a link to ritual and a meditation on the season.

Creating a wreath that is a labor of love is a reminder that no matter where we are in the seasons and phases of our lives, we will continue to cycle through the good, the bad, the ups and downs, youth and aging, life and death. Taking comfort in the recurrence

of winter, spring, summer, and fall. Drawing vitality from each to celebrate the mood and spirit of the season. Each of us, according to season or phase, will find ourselves moving along the wheel of life, marking the present, utilizing the greens and botanicals that are thriving at the time. Holly and ivy, rosemary and bay, spruce and pine to bring the best of the outdoors in when we celebrate solstice and Christmas in the bleak midwinter. Pussy willow and forsythia to herald spring. Lavender and roses to celebrate summer, and fruits, nuts, grains, and fall foliage to bid farewell to another growing season. Wreaths of laurel for times of celebration, baby's breath for birth, and acacia for times of mourning.

After the holidays, I move my wreaths to trees along the edge of the woods. I pay these fading seasonal markers homage by giving them back to nature and making way for the next turn of the wheel. Seemingly in tribute to this cyclical process, the fruits, seeds, and berries turn wreaths into neighborhood bird feeders, and the wildlife that replants them create new generations of growth as faded botanicals are sown into the surrounding landscape. Once the wreaths are picked over, they almost inevitably become places for nesting. Another visible reminder of the circle of life and our own creative contribution to the larger wheel of life.

WUTTAHIMNEASH

IN NATIVE AMERICAN HERBALISM, and all around the world, the doctrine of signatures, which related a plant's appearance to its properties, was a prescriptive memory aid for those who could not read and write. The heart shape of the strawberry not only gave it its name, wuttahimneash (literally "heart-berry" in the Algonquian languages), but it helped remind many indigenous peoples that strawberries are good for heart health.

The English name for strawberry is so deeply shrouded in the mists of time that ordinary speculation is hardly worth repeating; suffice it to say that "straw" runs through most etymologies. And technically speaking, strawberries are not even berries (berries have seeds on the inside, for one thing); they do, however, have springtime flowers that could make you blush, or a bumblebee bumble. Any chipmunk, catbird, or child can appreciate that they are the first berry of spring. The tender-sweet red fruits melt in the mouth, reminding us once again why eating seasonally is sublime; when ripe, they are laden with vitamin C, help to thin the blood, and put a red-lipped smile on our face. And the evergreen leaves have long been used in winter, a time when few other greens are available, to bring nourishment to stews.

You can choose *Fragaria
vesca* (woodland strawberry) if
you want an upright perennial
suitable for edging, knot gardens, or massed plantings; the
berries are thimble-sized and
sweet as can be. *Fragaria virginiana*, a wild strawberry from eastern North America, sends out
runners and yields larger berries
than *F. vesca*; as a vining perennial, it can simply be planted in
the garden or used in hanging
baskets, window boxes, or to
diversify your lawn. *Fragaria*

×*ananassa*, the modern domesticated garden strawberry, was first
bred in Brittany in the 1750s by crossing *F. virginiana* and *F. chiloensis*, which was brought from Chile in 1714; its cultivars have virtually
replaced the woodland strawberry in commercial production.

Strawberries are often grouped according to their flowering
habit. June-bearing plants produce their fruit in early summer, and
ever-bearing plants are just that: throughout the season a single
plant may produce fifty to sixty times, or roughly once every three
days. Heirloom strawberries are easy to grow as a cottage garden or
specialty you-pick farm crop, but they do ripen so "quick and tender" that they need to be eaten almost immediately. Few other foods
demonstrate so well the marked contrast between sun-ripened fruit
meant to be consumed fresh, and the chemical-laden (and frequently
flavorless) big ag versions, bred to be dense as ham for ease of shipping and maximum profit.

Throughout history, strawberries were most often enjoyed ripe from the plant or as table fruit. In addition to eating them fresh, many Native Americans dried them, or ate them in fritters. English and French settlers adopted this last preparation with gusto, as well as quickly coming to enjoy the seasonal abundance in simple bowls of strawberries and cream. As rhubarb increased in popularity, it was frequently blended with strawberries, which added sweetness to the tart rhubarb stem. By the 19th century, strawberry shortcake had become a national favorite, often associated with June strawberry festivals and fairs and Independence Day celebrations. In addition to these traditional desserts, strawberries were also preserved for winter and market use in the form of jam, jelly, fruit leather, and ice cream.

In the landscape, strawberries serve as a wonderful groundcover. They help with erosion control, and they make great native habitat plants, acting as a supplemental food source for wildlife and pollinators—and kids and families. In a layered and complex world, there is still always a place to enjoy an old-fashioned strawberry festival, buy a fresh pint, or eat your own sun-ripened berries. That can't help but make your heart feel better.

XERISCAPING

WHETHER WE LIVE IN A DESERT REGION, or we simply aspire to cultivate more sustainable gardens, xeriscaping enables us to enhance a full-sun site with drought-resistant plants that thrive without irrigation. Xeriscaping (from the Greek *xeros*, "dry") literally refers to working with the ecology of an environment or habitat that lacks moisture. It is learning the "art of arid," so we can fill even the drier parts of our yard with green. Ultimately it means planning for a water-wise landscape—something that we should probably have been doing all along.

In recent decades, "right plant, right place" has been the mantra of horticulturists everywhere. For too long, we forced lawns and foundation plantings to conform to the contractor's discount special: "green meatball" shrubs (as Michael Dirr calls them) on either side of the front door, rocketship arborvitaes at the house corners, and a few dots of color that grew into the foundation and over the windows within ten years. Since we started stripping building lots of all life and soil, most lawns around new homes required the installation of irrigation and a chemical regime to survive.

Right plant, right place appeals to the better wisdom of a gardener. It means understanding that large tender leaves often need shade, or a gluttonous share of water and nutrients to keep them

alive. It means knowing the mature
height and breadth of a plant, so
that they can hold their place without us torturing them with harsh
pruning, hedging, or transplanting, all of which lead to botanical
strife—and a greater need for irrigation and fertilizer. Right plant,
right place teaches that in arid
portions of our landscape, deep-
rooted plants, succulents, thorny
and bristly flora, and plants with
leathery leaves are better able to
survive; it's knowing that silver-gray leaves, felted or hoary leaves,
and waxy narrower needles not only tend to endure drought but also
better withstand wind and other harsh conditions.

Succulents rely on good drainage and can even seem to grow
in sand; but other drought-tolerant plants in our xeriscape, such as
stachys, yarrow, and mullein, rely heavily on mulch. Mulch certainly
helps to suppress weeds that rob nutrients from our intended plantings, but it can also release nutrients that benefit all life above and
below ground as it decomposes. Sea and sand seem to go well together,
so I especially love to use seaweed in more arid garden spaces.

Moving from worst to best xeriscape soils in my yard: prickly
pear cactus, yucca, euphorbia, thistle, sedum, and sempervivum
merge into plots or pots of Mediterranean herbs like thyme, rosemary, hyssop, sage, and lavender; next, I may have tufts of chives,
dill, artemisia, artichoke, and horseradish, and even adjacent beds of
root crops like carrots, parsnips, and sunchokes that can store moisture beneath the ground.

Every garden plot makes a difference, and every single tree we plant will absorb one ton of carbon dioxide over its lifetime. That's not small potatoes. Managing an arid landscape means planting a cedar or black walnut or other deciduous tree for shade's sake, and leaving their fallen leaves to decompose and make soil. Even fruiting trees and shrubs like blackberry, aronia, and shadbush can grow in more arid environs; their layers too help to provide shade and keep moisture in the ground. We can craft a rain garden or bioswale from a gutter spout, to guide pooling water whenever it can be had, or perhaps grow some iris, horsetail, agastache, or echinacea, to help a small glen flourish where a spent lawn may have struggled.

If we do choose to maintain lawn, a minimal and diverse lawn can grow parched and survive in a water-wise xeriscape—maybe even blossom into a meadow in wetter seasons. With a little encouragement, our own backyard can turn into a flowery mead made up of rooted ornamental grasses, milkweed, and tuberous plants like daylily, dahlia, canna, and crocosmia, which store moisture below ground and enliven our yards above ground with color, fragrance, swallows, and swallowtails.

Our xeriscape can be our best means to enhance the arid microcosm we are stewarding. With climate change, my region is losing fruit set from orchard trees that bloom too early, and it is predicted that we will likely lose maple syruping as cold-loving maples decline. To offset these losses, we can experiment with native trees and riparian plants from the next nearest zone to the south. Sometimes that might mean trialing new economic crops like our native pawpaw, or temperate fig and persimmon trees, which we were previously unable to grow.

Taking it a step further, we can apply the craft of xeriscaping to the pavement under foot and the roof over our head. We can convert sun-baked rooftops into absorbent green roofs and improve our lot, literally, by turning parking areas and driveway edges into permeable greenspaces that absorb runoff instead of requiring irrigation. I have even seen inspiring knot gardens and parterres that recreate old planting patterns with drought-tolerant succulents, helping us adapt to, enhance, and even embrace a changing world, much as we welcome the cardinals that have moved north into New England.

Xeriscaping does not mean that my yard looks like an upscale Albuquerque parking lot; it just means that my lawn fades back as water resource management plays into my landscape. And xeriscaping becomes first nature instead of a second thought. I encourage a stand of drought-tolerant native sumac to stabilize an eroding slope. I grow grapes to provide shade beneath an arbor, site espaliered peaches against a south-facing wall, and cultivate drought-tolerant plants everywhere they will thrive. I grow zucchini in my own compost pile instead of buying jet-watered agricultural product shipped thousands of miles from a ninety-acre field. I pull weeds before they compete for soil nutrients—and even weeding becomes an act of conservation. We don't have to know it all. We have things to learn together. Right plant, right place.

Often our hottest landscapes are along the street—which is fortunate, because front-yard conversations win back common ground. Environmental mitigation can be more easily won with tomatillos and garlic, or maybe even a lavender hedge. My plot to intercept climate change, and your plot down the street. The news is turned off, and I'm back out talking with a neighbor. We talk about what is working and what we need to do. I get mulberries. She gets mulberry liqueur or tomato sauce, depending on what came my way. Your milkweed seeds blow down the street, and together we cocreate a meaningful shift, making a greenbelt that helps offset the effects of drought and climate change.

YARDS

MANY HORTICULTURISTS MAKE a point of distinguishing garden from yard. Yard is the assemblage, our curated accumulation of sun and soil, productivity, and recreation. Our habitat. In the last century, we held ornamental gardens in higher esteem than edible landscapes, but we need both to nourish us. Sometimes we elevate the garden to the neglect of the environments that support it, choosing lawn over meadow and thicket, Asiatic lilies over lettuce, hostas over kale. But our yards can be reclaimed; we can have diverse woodland edge instead of a wooden fence, native shadbush and elderberry instead of forsythia and privet. In essence, a garden of our many parts, woven into our yard.

Our yard is our personal intersection with nature, carved out and arranged into work areas, cultivated spaces, circulation routes, and organized places to experience outdoor living. Many of us appreciate nature for its wild side, but gardens have long been esteemed as the highest achievement in our culturing of the land—the culture of horticulture as a means to elevate mind, body, and spirit; high art for our cultivated parcel. But the comprehensive skillset of a yard is just as worthy. It is a reflection of our ability to cocreate our own habitat—our countryside, suburban, or urban microcosm. Front yards, kitchen gardens, work and living spaces created with beauty

We abuse land because we regard it
as a commodity belonging to us.
When we see land as a community
to which we belong, we may begin
to use it with love and respect.

—ALDO LEOPOLD

and utility in mind. Intersections with nature and its wild side, bent toward home production, whether it consists of five pots on a stoop, a one-acre foodscape, or a family farm. If we are fortunate enough to have any space to spare, our yards enable us to turn a house lot into a vibrant homestead: a yard scaled to meet the needs of an individual, family, or shared community space.

Language often reminds us about the origins and arrangements of the spaces we inhabit, and digging into our yards can be telling. In Old English, *geard* (from the Proto-Germanic *gardan*) referred to a fenced garden and residence; over time, Middle English *yerd* evolves into yard. In Late Old English, *orceard* (from the even earlier *ort-geard*, "hort yard") referred to a fruit garden; this, of course, evolves into *orchyard*, our orchard, over time. From orchard to herb-yard, bee-yard, stable-yard, farm-yard, barn-yard, stock-yard, vine-yard— this is where we tended to our yardwork, this is the landscape where we exercise our craft.

Venturing down along the river where I live, I pass an old brick-yard, a mast-yard, and shipyard. A railyard, a schoolyard, and a churchyard—though here in old Maine, the churches don't have boneyard or kirkyard cemeteries. Instead, nearly every old house has its own graveyard—just another part of life and land management for old Yankees. But they planted their graveyards, too, and each year their patches of daylilies, lilacs, lilies of the valley, and bleeding hearts come back to life.

In my region, we still refer to our gardens and landscapes as our yard, something that can bring strange looks to people "from away." Yard refers to the patch of land around a house, an arcane English expression from a time when you might have been lucky to have a property to live upon and cultivate. Farmer-poet Thomas Tusser's *Five Hundred Points of Good Husbandry* (1573) spoke to ordering your yard in ways our grandparents would recognize: a yard to capture sun and shade and corresponding garden rooms to provide respite and nourishment. The features they interacted with most frequently were closest to the house, the less often used (and utilitarian) further distant, and the most elegant offerings streetside (or riverside, in an earlier age).

History and economics determined the balance of garden and work space within the yard. The dooryard garden was how you presented to the world—like the clothes you wear when you go out, an expression that you cared. The dooryard was a public interface, made engaging with ornate gardens, a place where neighbors could meet over the fence and where news, plants, fragrances, and conviviality could be shared. A work yard was positioned according to the needs of the household. Typically, it would include an open shaded space where outdoor work like laundry, plucking, butchering, soap-making, and related artisanal crafts could be carried out. In larger landscapes or urban settings, the courtyard offered an enclosed space to shelter and protect the people, plants, and belongings contained within.

Many colonial-era inventories, volumes like Gervase Markham's *The English Huswife* (1615), demonstrated best practices for laying out and ordering a property from doorway to distant field. Early American colonists did all they could to emulate English landowners; although few had owned property before arriving on these shores, many had farmed and carried out trades tied to the landscape previously, and they adapted their skills to suit North American landscapes. As time went on, Thomas Jefferson modeled best practices for a new nation; he traveled around Europe to learn more about the art and science of agriculture and then shared models suitable for New World backyard gardens and farmyards.

By 1869 and the First Transcontinental Railroad, much of our regional agriculture began shifting to the fertile Midwest, and new urban centers grew up around the railyards. Frank Scott, in his *The Art of Beautifying Suburban Home Grounds of Small Extent* (1870), urged these urban dwellers and the swelling ranks of former farmers to meet on middle ground. His models for the suburban yard—complete with lawn, garden furnishings, and ornamental gardens—were designed to celebrate the newfound leisure time and disposable income of the rising middle class. Farms were dissected into smaller yards, and the new residents of suburbia reassessed the amount of yard space needed to supplement household diet.

With the introduction of the automobile, many garden spaces, work yards, chicken yards, and laundry yards were given over to driveways, garages, and industrial-age work rooms for the outdoors. Earlier waste management shifted from privies and stable yards to indoor bathrooms and garages, where trash cans were parked alongside the token shovels, hoes, pots, and planters that moved in from disappearing barns. It wasn't until World War II that we reassessed what we had lost. Victory Gardens once more coaxed productivity from the land, and for many of us, these wartime yards helped shape our 21st-century perceptions of living in a fruitful community landscape. We are now in a similar time of reconsideration, adapting to an even more densely populated landscape. We are once again reminded that food security works best when there are many alternative sources for edibles—sources including our own backyard.

Today my homestead yard supports the intersections between ecosystems and my current interests. My pleasure gardens include flower, herb, and vegetable gardens. The surrounding landscape is beautified with favorite ornamental and edible trees, shrubs, perennials, water features, sculptures, a fire pit, and seating areas to enjoy the views. Over the years, my work yard has included a sugar bush, orchard, bake oven, and laundry line; woodpiles, compost piles, bees, small livestock or fowl, elder groves, blueberry barrens, and various spaces for project work. All help keep me productive and entertained at home; I don't need to go out and buy into the consumer culture. Something as simple as a laundry yard feels to me like mindfulness in action. Pinning up each garment to best capture the wind or sun. A mindless meditation where my mind drifts on breezes too.

We pay for experiences to relax our mind or enjoy the outdoors. My laundry does that weekly. My clothes last longer; they self-iron as they hang and smell fresh without toxic dryer sheets. It's not for everyone, but my laundry yard is also a part of my sustainable homestead. It helps me to be less reliant on systems that I do not want to support. I am not a purist, and I will use a dryer during rainy seasons, or when waiting on my line is not reasonable, but I also grew up knowing the statistic that we wouldn't need a single nuclear power plant in America if we didn't use clothes dryers. For me, designing a yard that enables the simple act of growing my own, or hanging out laundry in the open air, becomes an act of resistance. Perhaps even more than that, I enjoy conspiring with sun, wind, and nature. Simply being present with a length of taut rope, wooden pins, the clothes off my back, and the nature of the day.

Home landscapes, homestead gardens, and closed loop farmyards are all reminders of our daily participation with stewardship and sustainability. Gathering kindling, tending a flock, composting yard waste, watering garden beds from a rain barrel—all become a cooperative process in participation with nature. My yard is the place, however humble, that I get to manage my flow of work centered around place. It's where I make sense of the day. And in the course of my work, I am taking in beauty, getting exercise, weeding my garden, tossing weeds to hens, and gathering eggs for sustainable protein and seasonal veg for more place-based deliciousness. I scour my yard for windfall, build a mesmerizing fire, and prepare food that I helped to create by living in process. A process that is not mine to live every day, but one which gives me enormous satisfaction each time I do.

Our relationships with the land are reciprocal, and our yards can be home base for environmental stewardship. The world is broken, but I believe that our job is to fix it as much as we can—no more, no less. My yard is my starting point. The grounding place from which I refresh and nourish myself as well as a small chance to cultivate a model environment for my local ecoregion—our habitat and all those we share it with. My yard is my connection to the greater ebb and flow, and my entrée into a world where I can see, feel, and taste my contribution—and know that I am not helpless to effect change.

Our gardens can mend, and weave, and our landscape can rebuild habitats and communities. For centuries, the yard has been the soul of a household. A place to retreat. Take in fresh air. Engage with the seasons. My yard is the place where I can find something to celebrate in life almost daily. A place where my helping hands have helped the earth to spring to life; and a place where your hands can do the same. Inch by inch, row by row, yard by yard.

YARROW

PLANTS ARE STORYTELLERS that draw us into the natural world, and some of these botanical revelations can be formative. Once when I was little, I cut my leg while chasing a dragonfly through a nearby field, and in response, some old Yankee neighbors introduced me to yarrow. They matter-of-factly plucked some yarrow leaves and handed them to me to crush in my fingers and hold against the cut. I was amazed: the bleeding stopped immediately. I'd witnessed magic from the natural world. A veil lifted, and I saw that plants had properties that ran far deeper than fragrance, taste, and beauty. They had deeper, scientific roots—and like people, they had homelands, personalities, names, and ways of expressing themselves. Most importantly, I learned that plants could be my allies.

The generic name of *Achillea millefolium* honors Achilles, of Greek mythology, who used it in battle against [y]arrow wounds during the battle of Troy. The specific epithet translates to "thousand leaves," because each feathery leaf seems to be made up of a thousand smaller fronds. The most common name for yarrow in years past was nosebleed, clearly a reference to the plant's styptic benefit; it was also known as soldier's woundwort ("wort" is a word applied generally to any medicinal herb), sanguinary (pertaining to or staunching blood), thousand-seal (a millefolium—to close up wounds), and old man's pepper (a condiment for those in the know).

Besides the doctrine of signatures, Europeans ascribed to another way of understanding medicine, one that was akin to traditional Chinese medicine. This doctrine of humors (from the Latin for "liquid") was based on an ancient Greek theory concerning four bodily fluids that were thought to determine health and temperament: blood, phlegm, choler (yellow bile), and melancholy (black bile). Common household botany ascribed qualities and "temperatures" of each humor to every herb, vegetable, and foodstuff used for nutrition and wellness. Humoral theory formed the basis of western medicine and had tremendous influence up to the 19th century. Herbs of an "opposing humor" would be prescribed or prepared to alter the condition being treated. Thus, fever would be treated with the application or ingestion of cooling herbs and vegetables. Phlegm would be allayed by using hot, dry spices. Melancholy, a cool and dry ailment, was healed by warm and moist (sanguine) herbs like St. John's wort and borage. And yarrow, being cool and dry, would stop the flow of warm, moist blood on a young kid's leg, hundreds of years from those ancient roots. It's also why yarrow tea was sipped for internal bleeding and to strengthen the intestines, liver, and kidneys.

Cool and dry herbs carry a bitter or tart flavor profile; no surprise, then, that the flowers, herbaceous leaves, and intensely bitter roots of yarrow have long been used in making liqueurs, tinctures, and digestive bitters that help to process fatty (warm/moist) foods. Historical culinary descriptors can often sound unpleasant, especially when marketing has caused most of us to equate "bitter" with "bad," but remember: many of us begin each day with a bitter brew known as coffee, sip a tart lemonade with lunch, or finish it out with a hoppy beer! In the Middle Ages, yarrow was a common part of the botanical blend lending a bittering agent to gruit; the same principle applied centuries later when we took to mixing bitters and tonics into cocktails for added depth of flavor—and to assist the kidneys even as we taxed them by drinking booze.

And now that we can determine the chemical constituents present in plants, we can understand why, in this instance, yarrow works as a styptic, astringent, diaphoretic, and tonic: chemical analysis shows that yarrow contains salicylic acid, isovaleric acid, asparagine, sterols, and flavonoids; and knowing this lineup helps explain and support many of yarrow's earlier uses in traditional or folk medicine and other ethnobotanical purposes. For example, yarrow's dark blue essential oil, made by steam-distilling the flowers, kills the larvae of the mosquito *Aedes albopictus*. A common branch of science also shows how several cavity-nesting birds, like the common starling, use yarrow to line their nests, suggesting that adding yarrow to nests inhibits the growth of parasites. Once again, we are reminded that we live in an interconnected web, and that there is benefit to looking for the commonalities we share when living together in habitat.

Yarrow is an especially useful companion plant: it repels some pest insects and attracts beneficial ones. When it flowers, yarrow attracts the good ones (ladybugs, hoverflies, predatory wasps), which drink the nectar and then hoover up the pest insects as food for their larvae. And it's easy to encourage yarrow in the landscape strictly as an ornamental perennial: it's fairly indestructible, spreads readily, and attracts butterflies and bees. The flowers are not the showiest (except upon closer inspection), but it brings a mass of color and life to garden beds and meadow gardens. If you are not enamored of the flowers, its feathery leaves are drought tolerant and mower-deck height, so it can be used like a groundcover to diversify sunny or dry-shade lawns.

Any good gardener knows that our gardens encourage us to keep our hands in the earth and grow alongside them. Plant allies like yarrow can even help feather your own nest, and certainly, one of the longest standing craft traditions in my life is with this old ally. Each spring, I gather the lush emerging leaves and flowers, and craft them into an oil that will be blended into my annual batch of herbal healing salve. Each batch carries lessons and reminders for me. One lesson from the field where we first met, one from my neighbors, one from old herbals, and one from yarrow itself. I can almost hear its cool, dry, raspy voice reminding me to harvest the healthiest leaves when they are at their peak. Later in the season, yarrow and some bees conspire to remind me that it's time to harvest the nectar-laden flowers. The result is a melding of natural world and a legacy of understanding, from ancients through scientific monographs, blended together in an apothecary jar.

ZINNIA

BOLD, BRIGHT, AND SUNNY, like a too-brash friend. If we didn't love zinnias so much, we might think them too loud for any civilized gathering. In a northern climate, the cacophony of bright colors can sometimes seem garish, though they meld in beautifully with the brighter colors of southern climates, where they are native. And then there are the hybridized Victorian purple doubles with mounded florets that look like a bumblebee orgy. Perhaps that is why bees and butterflies love them so. As for me, I am a fan of the heirloom Red Spider zinnia (*Zinnia tenuifolia*), which punctuates borders with single scarlet starbursts everywhere it manages to creep through the garden.

Zinnias were first encountered by the Spanish in Mexico and brought to Europe in the 1500s. *Zinnia peruviana*, a rangy species with small spidery flowers, was known in Europe by the mid-1600s, but it was not widely cultivated. Johann Gottfried Zinn of Gottingen University in Germany wrote the first description of the plant in the mid-18th century. In 1796, another zinnia was brought to Europe, possibly by way of Brazil, and presented to Linnaeus, who named the entire genus in Zinn's honor. This *Z. elegans* is the ancestral plant from which our modern zinnias have been developed; it produced larger flowers in a wider range of colors. Europe and Great Britain

further cultivated double-flowered zinnias, and by 1864 zinnias in a broad spectrum of colors had made their way back to North America, where they were right at home in the flamboyant carpet beds that were fashionable in Victorian-era gardens. No garden then would have looked right without zinnias, and no cutting garden now would be complete without these long-stemmed, showstopping flowers.

When we want to engage children in gardening, zinnias are great place to start; they are easy to grow from seed, and the plants are rugged stalwarts of the garden, flashy enough to gain a kid's attention or the notice of any passerby. As much as I liked them as a child, I forgot to include them in my grown-up gardens for years. These days in a more challenging world, I find that they gift us with bountiful, bodacious cut flowers, cheer that we can share—and the reminder that there is always a place for a raucous friend who keeps us smiling.

ZUCCHINI

ZUCCHINI IS THE EPITOME of seasonal bounty in a green rind. The vines yield gratifying horticultural and culinary results from seed to table, producing harvests so abundant, they can't help but foster a spirit of generosity. And they are a fun plant to encourage any novice, making any child or amateur gardener feel like a master gardener. The seeds, shaped like squirrel eyes, are particularly easy for kids to handle; they germinate in as little as a week, so they tend to hold the focus of kid gardeners with short attention spans and the need for quick success.

My compost pile has taught me that the best squash tends to grow in rotted manure; so, I discovered two lessons there. The first is that zucchini likes to grow in hills of rich soil, especially the vining types. The early crops that sprout from seeds in my compost pile remind me of the second lesson: compost runs hot, and it's worth directly sowing seeds into rich compost to get a significant head start on the growing season. By hilling them in a garden, gravity and the vines can expand the reach of flowers and fruit, while roots take up endless nourishment from the rich soil below.

Fried or stuffed zucchini flowers are an Italian delicacy, and since vines flower so quickly, they offer some of the first garden edibles each summer. Male flowers are the first to appear on zucchini plants, opening before female flowers. Male blossoms have a long, thin stem, and they often open, release pollen, and then drop. Instead of a stem, female flowers have a swollen ovary that develops into the fruit; they must remain on the plant until pollination is successful and the zucchini starts to grow.

All pollinators seem to love the flowers, but native bumblebees seem most at home inside them, lolling about, dusted in rich yellow pollen from antenna to their bee's knees. However, I love the flowers too, and since the female flowers yield zucchini, I harvest only the male flowers. Sometimes I just add the bright golden orange petals into a quiche, frittata, or salad. When the season offers a bountiful surplus, I either tempura fry them or stuff the blossoms with a ricotta mixture, goat cheese, or soft feta and herb blend, dredge them in flour, and pan fry for early summer garden treats.

Zucchini and other summer squash represent the abundance to come, but they arrive at a time when we are still in disbelief that a little garden plot can yield so much. So much, in fact, that where I live, friends and neighbors regularly drop them off on each other's doorsteps, mailboxes, and unlocked cars. It can become something of a joke, but the surplus has been a plus as long as they have been grown.

Zucchini, a summer squash of Mesoamerican origin, was originally brought to Europe in the 16th century; the English called the new vegetable marrow and the French, courgette. The green selections we know as zucchini were developed in 19th-century Italy (their name is the plural diminutive of the Italian for squash, *zucca*). Ironically, even though summer squash originated in the Americas, zucchini did not appear in North America until it was brought by Italian immigrants like my grandparents in the early 20th century.

In my family, the larger family-sized zucchini were most commonly stuffed and baked with whatever was at hand—breadcrumbs, herbs and cheese, or herbs, rice, and meat. Most summer nights, zucchini would simply be sliced up and sautéed in good olive oil, minced garlic, and finished with some chopped herbs and tomatoes. When later summer vegetables like peppers, onions, and eggplant came in, the French influence broadened our palate, and ratatouille became a staple of late-season tables. It's a hearty vegetable dish that I never tire of, a healthy alternative when "gifted" drive-by zucchini donations—and a reminder that all good things come back around in time.

Index

F

fairies, 191
fairy houses, 91
farmers markets, 21, 92, 96, 117, 145,
        167, 194
farm stands, 21, 120, 145
female tonics, 206
feminine energy, 92
fences, 13, 26, 50, 52, 61, 69, 71, 101,
        134, 214, 216–219, 230, 232, 233
fennel, 73, 74, 94, 196
fermentation, 113
fertilizers, 40, 42, 124–126, 157, 167
fiddleheads, 45, 86–88, 96, 206
figs, 94, 228
fireflies, 139, 140
fish, 74, 194
flavonoids, 240
flavorings, 65–66
fledglings, 49
floral butters, 91
floral ice cubes, 92
floral water, 65
floriculture, 89–92
floriography, 138
flowering quince, 173
flower petal mandalas, 92
flowers, 135–138, 153–154
foliage perennials, 182
folk medicine, 240
folk traditions, 138
food safety, 186
foraging, 93–96, 104, 112–113
forget-me-not, 136
forsythia, 222, 230
fox, 135

foxglove, 71
Fragaria ×ananassa, 224
Fragaria chiloensis, 224
Fragaria vesca, 224
Fragaria virginiana, 224
fragrances, 64, 84, 113, 233
Francis of Assisi, Saint, 49
Frank Jones Brewery, 127
French sorrel, 192
fritters, 61, 74, 75
frogs, 53
fruit, 29, 36–37, 55, 56, 71, 93, 94, 102,
        104, 169, 191, 222, 228
fruit salad, 72
fruit soups, 18
Fuller, Buckminster, 11
fungal disease, 34
fungus, 52, 80
Funkia, 184

G

Galium verum, 92
Gandhi, Mohandas K., 97
garden craft, 11, 97–101
garden furniture, 60
Gardner, Isabella Stewart, 154
garlands, 91
garlic, 61, 85, 247
garlic mustard, 95
geard, 232
genetic diversity, 35–36, 41, 80, 106
gentian, 71
Gerard, John, 209
germander, 64
Germany, 102
gig economy, 24–26, 98–100, 145
gin, 57

About the Author

RACHAEL MONTEJO

JOHN FORTI is an award-winning heirloom specialist, garden historian, ethnobotanist, garden writer, and local foods advocate. He is executive director of Bedrock Gardens, an artist-inspired public sculpture garden and landscape in Lee, New Hampshire, and the recipient of a 2020 Award of Excellence from National Garden Clubs. He is also a regional governor and biodiversity specialist for Slow Food USA, a national chapter of Slow Food, a global organization and international grassroots movement that connects food producers and consumers to champion local agriculture, farmers markets, and traditional, regional cuisine. John gardens and lives along the banks of the Piscataqua River in Maine. Visit him, The Heirloom Gardener, at jforti.com.